James Robinson Nichols

Chemistry of the Farm and the Sea

With other Familiar Chemical Essays

James Robinson Nichols

Chemistry of the Farm and the Sea
With other Familiar Chemical Essays

ISBN/EAN: 9783744715430

Printed in Europe, USA, Canada, Australia, Japan

Cover: Foto ©berggeist007 / pixelio.de

More available books at **www.hansebooks.com**

CHEMISTRY

OF THE

FARM AND THE SEA.

WITH OTHER

FAMILIAR CHEMICAL ESSAYS.

BY

JAS. R. NICHOLS, M.D.,

EDITOR "BOSTON JOURNAL OF CHEMISTRY," MEMBER OF
MASS. INSTITUTE OF TECHNOLOGY, ETC.

BOSTON:
A. WILLIAMS & CO., 100 WASHINGTON ST.
1867.

Entered, according to Act of Congress, in the year 1867, by

A. WILLIAMS & CO.,

In the Clerk's Office of the District Court of the District of Massachusetts.

Stereotyped at the Boston Stereotype Foundry,
No. 4 Spring Lane.

PREFATORY NOTE.

THE brief chemical essays contained in this volume are presented substantially as they have appeared in the form of addresses before agricultural and scientific bodies, and as contributions to the *Boston Journal of Chemistry*, during the years 1866 and 1867. The unusual favor with which they were received by large numbers of listeners and readers, has led to their reappearance in the more convenient and permanent form of a book. The aim has been to present scientific facts and principles in a familiar way, so as to interest and instruct those not specially acquainted with matters of science, and if the essays prove acceptable and useful to the new class of readers to whom they are now introduced, the end of their publication will be fully reached.

J. R. N.

LABORATORY, 150 CONGRESS STREET,
    BOSTON, May, 1867.

CONTENTS.

	PAGE
CHEMISTRY OF THE FARM,	7
CHEMISTRY OF THE SEA,	47
CHEMISTRY OF A BOWL OF MILK,	55
CHEMISTRY OF THE DWELLING,	63
CHEMISTRY OF A KERNEL OF CORN,	85
OBSCURE SOURCES OF DISEASE,	91
LOCAL DECOMPOSITION IN LEAD AQUEDUCT PIPES,	97
BREAD AND BREAD-MAKING,	103
CHEMISTRY OF THE SUN,	111

CHEMISTRY OF THE FARM.

IN considering the Chemistry of the Farm, we must, first of all, bring to notice that accumulation of wonderful and important facts, which unfolds the philosophy of the origin, the structure, and the growth of plants. In darkness intense as midnight was this knowledge involved for centuries, and it was only by the light of those fires in which were buried the crucibles of the chemist, that the dark cloud was pierced, and all around and beneath illuminated.

The germination and growth of a plant is strictly a chemical problem, and intimate indeed is the connection of the soil cultivator with its perfect development. He has not the power to compel the aggregation of atoms; the unseen Manipulator whom we designate the "Vital Force" is the chemist who performs this marvellous work, and whose skill far exceeds all human capability. His laboratory is no circumscribed one, bounded by partitions of wood and stone, but its area extends farther than the eye can reach, and its enclosing wall is the great rotunda whose span stretches beyond even the imagination of men.

The farmer labors within this great rotunda, and in the immediate presence of the great Chemist, who invites him to aid in his work. Day by day he witnesses his marvellous power, in calling from the slumbering earth the tender blade of grass, the beautiful flower, the useful cereal and leguminous plants, the creeping vine, and the spreading oak of the forest.

He can promote or destroy the work of the great Creator and Architect; he can retard or facilitate the chemical changes which are going on so continuously and vigorously around and beneath his feet.

And what are these changes? A knowledge of them teaches the great secret of plant growth. It unfolds the philosophy of that fact, incomprehensible to so many, how, from the ethereal atmosphere, almost alone, the solid forms of organized structures are elaborated.

How wonderful is the fact, that a large proportion of the material of the grains, and fruits, and grasses which we gather into our barns and granaries, is composed of the constituents of common air! Perhaps it is even more wonderful, that the solid and inflexible fibres of the oak, the hickory, the beech, and scores of other woods, exceeding even these in density and hardness, are formed from the unstable medium we breathe, and which seems so utterly devoid of materiality and solidity.

Chemistry alone is capable of teaching us the philosophy of that aggregation of atoms by which plant organisms are developed and increased, until full maturity is attained. It teaches us respecting the office the soil, the rain, the air subserves in accomplishing the work; and the information it furnishes is minute, wonderfully exact, and full of interest to the student. It teaches the interesting fact, that the soil originates from the solid rock which constitutes the crust of the earth, and explains the nature of the forces which have produced crumbling and decay in the same. Its teachings are so important in this particular, that we will stop a moment to consider them.

If we procure from one of our hills a piece of granite of either of the different varieties, and finely pulverize and analyze it, we shall find it to contain all the constituent elements of which all other rocks consist. Hence we shall be led to conclude, that they all originate from the granite; that this is the parent rock of quartz, talc, serpentine, feldspar, mica, &c., from the crumbling of which our soils have been formed. By the decomposition and crumbling of the mica and feldspar in a particular region, one kind of soil is formed; by limestone in other localities, another kind; and hence it is plain to see that a variety of soils must result from the disintegration of the different kinds of rocks. A very clear conception of the work of exfoliation may be obtained by supposing an individual to have been placed upon our planet at a time when it was a hard, impenetrable mass of rock. Suppose him to have lived through all the great epochs of time until the present, and to have witnessed the gradual metamorphosis from barren sterility to the extreme of vegetable luxuriance. Suppose him capable of witnessing the gradual crumbling of the adamantine masses, and the formation of cultivatable soils. If the agencies in past ages were the same as are now at work, he would have seen that every flash of lightning shooting athwart the sky, by decomposing the atmosphere, produced a trace of nitric acid, and that this, falling upon the rocks, aided in the work of separation. He would have seen that the carbonic acid of the air, the rapid freezing and thawing, the mechanical effects of rain, the attrition of dust moved by winds, all conspired to reduce the seemingly defiant quartz, and talc, and gneiss, to a finely subdivided powder capable of sustaining vegetable life. The chemistry of these atoms of dust is very easily understood. The Creator, in the beginning, made use of about sixty different kinds of materials in constructing our planet, and he selected only ten or twelve of these from which to form all kinds of rocks. It follows that the dust atoms must be made up of the same materials as the parent rock. From them the mineral food of plants is obtained. The inorganic or mineral food which plants require are principally silica, lime, magnesia, sulphur, potash, and soda. Their presence in the soil is indispensable, as without them no plant growth could begin and continue. A plant has as capricious an appetite for its mineral food as a human being has for his food, and each variety calls for its appropriate nutriment; and if nature does not supply it sufficiently in the soil, or if we do not step in and furnish it, it famishes and dies. There is as much propriety in saying, when we observe a stalk of corn struggling for existence in an impoverished soil, that it is starving to death, as there is in saying that an animal famishes when food is withheld. Let us observe still further the striking analogy between plant life and animal life. I have said that both have their appropriate, chosen food. If we place before a cow or horse some forms of food which man requires, and withhold hay and grain, they will ultimately perish. Thus it is with vegetables. If we plant peas or beans upon a field where no trace of lime is found in the soil, although it may be rich in minerals which other plants would live and thrive upon, they will as certainly famish as though we sowed them in the granite quarries of Quincy, or among the glaciers of the Alps. To attempt to feed the different varieties of plants upon the dust atoms of a single kind of rock, would be as absurd as to gather the different races of men together, and endeavor to sustain them upon the watery fruits of the tropics. While the seething negro would satiate his appetite, and grow lusty, upon the watermelon and the banana, the greasy Esquimaux would cry aloud for his train oil and blubber; and if withheld, he would probably die from the cravings of unappeased hunger. A plant is like an infant, as respects the preparation of its food. It has no teeth to masticate, no salivary glands to pour out diluting fluids, to render digestible its rocky aliment, and yet it can receive it only in a liquid, soluble form. Its mouths are microscopic, and nothing not minutely subdivided can pass their portals.

Farmers are men nurses, laboring among their plant children, pulverizing and moistening their food, just as the female nurse, within the precincts of the children's nursery, is busily employed in preparing and rendering easily digestible that which the appetite of her little troop so urgently demands.

Nature does much, by the activity of those forces already alluded to, in preparing the inorganic food of vegetables. Although the rocks have crumbled into powder of varied fineness, and the mass of this constitutes the soil, yet the largest portion is still very far from being fine enough to be appropriated by plants. Minute atoms of granite, of limestone and feldspar, scarcely perceptible without the aid of the microscope, pervade every soil, and must be further acted upon by carbonic acid from the air, by rain, by mechanical forces, &c., before they are of any use to our maize plants, tubers, grains, or vines.

It will be clearly understood, that we may possess land rich in the mineral substances which a particular grain requires, and yet, after successive crops, it may languish and fail for the want of a substance already in the soil, but which is not in a condition to be used by the grain. In this we see the connection of chemistry with the business of the farmer in the tillage of his lands. He plies vigorously the plough, the hoe, and the cultivator; he digs, he pulverizes, he reverses the condition of the soil; bringing up to the surface that which was buried, and burying that which was upon the surface; and does he suppose that the vigor he thereby imparts to the soil and plants is due solely to the mechanical effects of his labors? There are great benefits thus produced which are far from being mechanical. It is indeed beneficial to loosen the soil so as to prevent binding, and to aid in the percolation of water through it; but some of the greatest benefits of active tillage are strictly chemical in their nature. By stirring the soil, atmospheric air is let in; and the carbonic acid it contains fixes its corrosive teeth into those minute grains of rock, and rends them asunder.

They are thus so changed, that, instead of being rejected by the hungry plants, they are seized with avidity, and consumed. And, further, by tillage there are chemical effects produced in that part of the soil not mineral or inorganic, by which decay or putrefactive change is carried forward, and plant food produced in large quantities. Thus chemistry conclusively shows that, by mechanical labor alone upon a soil, nutriment is afforded which is equivalent to the application of manure; and with these facts distinctly in mind, the farmer need not be surprised at the energy with which his crops shoot forward after the application of the hoe and cultivator.

It was chemistry that taught the husbandman the importance of subsoil ploughing. There are many farmers who are unable to overcome their prejudices sufficiently to try the experiment of deep ploughing upon their soils. They suppose the whole virtue of their lands lies in the black mould or humus upon the surface; and if they go below, and bring up sand, and yellow or pale earth, and mingle with it, of course it must dilute and impair its fertility. They certainly know that their soils are superficial and weak enough, without going down to bring up that which cannot sustain, as they suppose, a blade of grass. They reason thus because chemistry has not taught them its important lessons. How important to remember that that which lies deep below the mould came from the rocks, and is rich oftentimes in their mineral constituents. It needs only to be brought up to the surface, so that air and rain can reach it, to promote chemical decomposition, and fit it for important plant aliment.

Chemistry teaches that plants do not obtain all the

elements of their growth from the mingled rock dust and humus constituting the soil. The atmosphere comes in for a share in rearing the structure, and the aid it renders is voluntary, and entirely independent of help from the husbandman. He cannot promote his interests and increase his crops by endeavors to influence atmospheric action upon his plants. It is only through the soil that he is able to do this. Plants derive their carbon, or charcoal, chiefly from the air. The great bulk of all plants is carbon, and consequently we see how important is the aid derived from that source.

How few of us call to mind the fact, as we sit around our comfortable hearth-stones in the long evenings of winter, and witness the gradual transmutation of the blazing pile of wood into black lustrous charcoal, and then, by further combustion, apparently into a heap of ashes, that there is in the one a constituent of the very winds from which we are so effectually sheltered, and in the other a portion of the soil abstracted from our fields. I am perplexed to understand how any one can witness these wonderful changes from day to day, and not have sufficient curiosity awakened to be led to interrogate that beautiful science which is competent to answer every question and solve all difficulties.

The facts as stated are certainly paradoxical and difficult of apprehension. There is no charcoal in the earth, none in the air; and yet, if we allow fire to act upon a bit of the whitest wood, or a grain of wheat, or corn, an apple, or starch, or sugar, it is always produced. Does fire produce it, manufacture it; or does it simply develop what was positively in these substances before? Chemistry affords an answer to the question. Suppose a good housewife places in her heated oven an apple (a potato, a loaf of bread, or any vegetable substance will serve the same purpose), and then, amid the multiplicity of household cares, it is forgotten, and when examined is found done, not *brown*, but *black*. The oven has inadvertently acted the part of a charcoal manufactory. The apple has disappeared, and in its place is found a dark and crispy shell. In its growth, the apple took from the earth its gaseous elements, its hydrogen, oxygen, and also its mineral rock food. From the air principally it procured its carbon, in the form of carbonic acid, which is a gaseous acid composed of one atom of carbon united to two of oxygen. Thus united to oxygen, it exists in the air, and although itself always intensely black, except when in a crystallized state, its color is not detected by the eye. We may perhaps be led to conclude that the apple, in common with all other vegetable substances, is ashamed of the color of its carbon ingredient; for before it can appropriate it to itself, it must first expel its two oxygen attendants, and thus expose its hue; but it instantly so blends and combines it with the other elements, that we are unable to see it until that merciless disorganizer, heat, drives off again its more fickle and volatile companions, and then the sable element is seen in all its nakedness. The undue heat of the oven has done this. While the oxygen, hydrogen, and nitrogen ingloriously fled, as the flame curled around the iron dome, black carbon remained faithful to his post. But let us try his courage a little further; let us see what curious results will follow if we apply flame to the crispy mass. And now we see changes and new combinations wonderful to behold. One of the substances, oxygen, which fled so precipitately from the oven, now seems to repent of its inconstancy; and as the flame grows more intense, it rushes into the very centre of the conflict, not singly, atom by atom, but in pairs, two individual atoms together, clasping one of the carbon; and thus the sable bride, again married, not to one oxygen bridegroom, but to two, floats off upon its bridal tour through the air. But such unnatural unions must always prove bad, and of short duration; such is the result here. The united parties are acid from the start. Thus combined, they constitute, in fact, carbonic acid, and the unhappy union continues until some beautiful plant, or flower, in seeming pity for the parties, seizes them in its tiny embrace, and with one strong effort effects their separation, sending the disunited atoms of oxygen away into space, and appropriating the carbon to itself, to aid in its extension and growth. When the charcoal is burned away there remains a small quantity of ashes, the mineral food of the apple derived from the earth. We venture to adopt this method to illustrate some of the marvellous changes incident to the growth and destruction of all vegetable organisms. Chemistry has taught us fully respecting these transmutations and the whole philosophy of plant growth; but let us pass to consider the chemistry of artificial fertilization.

The dark heaps of animal excrement which lie about the barn-yards of farmers, have, during all ages, been known to possess specific fertilizing influence upon plants; and if it were furnished in sufficient quantities to replace the elements removed from soils in repeated croppings, the labors of chemists in the direction of seeking out new supplies of plant food would be practically aimless and absurd. But this is not the case. The exhaustive process is continuous in all civilized communities, and it is impossible, in densely-peopled sections, to maintain a satisfactory balance between supply and demand.

It was very natural, then, that early in the history of chemistry as an exact science, it should be called to the investigation and determination of the chemical nature of that material, which common observation and experience had taught to possess the natural food of plants. As regards its superlative value, no one has ever entertained a doubt, either before or since the field of chemical investigation was fairly opened. What is its composition? Allow me to present the results of some determinations of my own on this point. A parcel obtained from the yard of a neighbor, which, under the conditions in which it was produced and preserved, may be regarded as a fair representative of the article as furnished by ordinary farmers, gave the following results: A portion weighing 7,280 grains was carefully dried in a porcelain dish over a water-bath, and it was found to lose of water 5,960 grains, leaving of dry matter 1,320 grains. Of the residuum thus freed from moisture, 455 grains were placed in a platinum capsule and carefully ignited, thus removing the combustible or carbonaceous matter made up of the elements — oxygen, hydrogen, and carbon. The resultant ash weighed 177 grains, showing a loss of volatile or combustible elements amounting to 278 grains. In order that the results of the analysis may be clearly understood, it may be desirable to present them without regard to fractional parts, and to estimate by the whole amount experimented with, viz., 7,280 grains. This amount gave of water, 5,960 grains; combustible or carbonaceous matter, 806; nitrogen, 29; potash and soda, 41; lime, 43; magnesia, 14; phosphoric acid, 15; sulphuric acid, 11; chlorine, 14; silicon or sand, 335; oxide of iron and alumina, 22. The points in this examination which will doubtless appear most striking, are the large amounts of worthless material which constitute the bulk of barn-yard manure, the water and sand greatly predominating over everything else.

A better idea of this may be obtained if the results of the analysis are applied to a larger amount of manure, which will give the constituents in pounds. Assuming that a cord of ordinary barn-yard manure will weigh three thousand pounds, its actual value as a fertilizer may be presented as follows: There is contained in it of water, two thousand four hundred and fifty-six pounds; common sand, one hundred and thirty-eight pounds. These added together give two thousand five hundred and ninety-four pounds of perfectly worthless substances. Now, if we still further subtract the carbonaceous matter, three hundred and thirty-two pounds, which is of no more value than muck, peat, straw or chaff, we have left only seventy-four pounds of active fertilizing material which has a money value. To obtain this seventy-four pounds, which really is all that is valuable, the farmer loads and hauls upon his field three thousand pounds, or one and a half tons, of a compound in which there is water enough to do the weekly washing of a small neighborhood, and a sufficiency of sand to keep the kitchen floor tidy for a month. The seventy-four pounds of mineral salts might be taken in an ordinary bushel-basket, and carried upon the shoulder to any point desired. In this amount there is the nitrogen, potash, soda, lime, magnesia, phosphoric acid, sulphuric acid, chlorine, iron, and alum. In estimating the market value of these substances, we may obtain the nitrogen by the use of crude nitrate of soda or sulphate of ammonia, at a cost of two dollars and sixty cents; the potash, soda, &c., in one and one half bushels of good wood ashes, at thirty-five cents; and fifteen pounds of common salt, ten pounds of bone-dust, three pounds of gypsum will supply the remaining constituents, at a cost of fifty cents. If we estimate the carbonaceous matter at ten cents, we have, as the actual cash value of all that promotes plant growth in three thousand pounds of barn-yard manure, the sum of three dollars and thirty-five cents. There are but few localities where the farmer can purchase manure at less than seven dollars the cord; and when to this we add the expense of hauling and applying to fields, we find there is a wide margin between the cost of the isolated valuable constituents of manure, and the article as furnished in its natural condition. Barn-yard manure may be imitated by thoroughly composting with a cord of seasoned meadow muck sixty-five pounds of crude nitrate of soda, two bushels of wood ashes, one peck of common salt, ten pounds of fine bone meal, two quarts of plaster, and ten pounds of epsom salts. The cost of this compost will not be over three dollars and fifty cents the cord, and ought, other things being equal, to serve as good purpose in the field. In practical trials of this mixture I have found that while it serves a most admirable end, giving very satisfactory results, it does not act so rapidly and energetically as manure; but its effects are more lasting.

In short, the same salts and organic matter as found in the dung-heap, have a higher money value, and seem to exert a more specific influence upon plants, than when presented in artificial mixtures. By substituting nitrate of potassa, or saltpetre, for soda, the compost is greatly improved, while its cost is enhanced. If the salts are dissolved in water, — those that are soluble, — and the bone in ley, and good muck is employed, a compost is formed very nearly as valuable as seasoned excrement. Very nearly, we have said — why is it not of equal value?

We have reason to believe it is owing to a minuteness of the subdivision of atoms, which we can neither produce nor comprehend, — a degree of comminution which sets at defiance all mechanical and chemical manipulation. Besides this, there is, however, a peculiar condition arising from, or communicated by, the contact of vital forces, which science is incapable of explaining. A physician once brought to me a jar of ox's blood, with the request that I would extract or isolate the metal iron therefrom, and place it in his hands. In answer to an inquiry regarding its uses, he stated he wished to employ it as a therapeutic agent, under the impression that iron once assimilated would have a higher and more natural influence when passed again through the animal economy, than the usual forms of the metal from other sources. His hypothesis was undoubtedly correct, and while it was quite within the power of chemistry to isolate the iron from the blood, it was impossible to secure it in the *condition* in which it existed in that fluid. That condition is indeed a peculiar one, and its presence is not indicated by any of the usual chemical re-agents. If we applied to it simply the usual manipulating processes, chemistry would fail to show that there was an atom of iron present in the blood of men or animals. This may illustrate the difference between the fertilizing influence of metals and salts, as found in animal excrement and as existing in other, or the usual forms. The iron as found in the blood, if administered to an enæmic patient, would without doubt immediately, and by direct and easy processes, again pass to its appropriate place, and restore the sanguineous fluid promptly to its normal condition.

But chemistry can never furnish it in that form, neither can it supply the mineral constituents required by plants, precisely as found in manures; but this must not lead us to disparage science and reject its teachings. We will accept what it does teach with sincere thankfulness. We will use as a medicine the best forms of iron it suggests, and they are many and of great efficacy; we will employ those fertilizing agents which it has pointed out as possessed of merit; and they, also, are many.

The impression entertained by some that chemists underrate and disparage barn-yard manure, is an erroneous one. It has no foundation in fact. They labor to multiply sources of this material, and the most im portant service rendered by it to the farmer is in the methods it points out whereby it is economized, and its efficacy preserved. In this particular, chemistry has accomplished much for agriculture. Would that soil-cultivators gave heed to its suggestions; then, indeed, would there be less demand for other agents.

But, secondly, let us consider what it has done in the way of furnishing a supply of these. Here we find the evidences of the exercise of a wonderful intelligence and industry,— a persistent scientific labor hardly excelled in any other field of research. It has analyzed and demonstrated the great value of decayed vegetable matter, as peat or muck; and given reliable directions how to fit it for manurial uses. There is scarcely a substance upon the land or in the sea that has not been made the subject of careful examination, with the view of ascertaining if it contained those principles capable of nourishing plants. As the results of these labors, we have a class of substances which, in contradistinction from animal excrement, or barn-yard manure, are called "special" or "chemical" fertilizers. Perhaps no article of the class has received more attention in this country and in Europe, than bones, and they have become a standard article of commerce. They are presented in the natural condition, as found in animals, or in that of a powder of variable fineness. Dissolved in acids, before or after calcination, they are called "superphosphates," and in this form are largely employed in agriculture. The term "superphosphate" is a popular one, and advantage is taken of this to palm off upon unsuspecting farmers all conceivable compounds of meadow-muck, human excrement, blubber and fish oil, gypsum and charcoal, as the genuine article.

Superphosphate of lime, or that compound formed by dissolving finely-ground bones in sulphuric acid, is a most excellent fertilizer. There is scarcely any land in New England that will not, under its use, render highly remunerative returns; but we cannot depend upon manufacturers for it. Every farmer must make it upon his own premises; and I insist that it can be produced readily, safely, cheaply. Let me present the method which I adopt upon my own farm premises.

Take a common sound molasses cask, divide in the middle with a saw, into one half of this place half a barrel of *finely*-ground bone, and moisten it with two buckets of water, using a hoe in mixing. Have ready a carboy of oil of vitriol, and a stone pitcher holding one gallon. Turn out this full of the acid, and gradually add it to the bone, constantly stirring. As soon as effervescence subsides, fill it again with acid, and add as before; allow it to remain over night, and in the morning repeat the operation, adding two more gallons of acid. When the mass is quiet, add about two gallons more of water, and then gradually mix the remaining half barrel of bone, and allow it to rest. The next day it may be spread upon a floor, where it will dry speedily if the weather is warm. A barrel of good loam may be mixed with it in drying. It may be beaten fine with a mallet, or ground in a plaster mill. If several casks are used, two men can prepare a ton of excellent superphosphate, after this method, in a day's time. It affords a prompt fertilizing influence, especially upon root crops, even when employed alone. Much less acid is used in this formula than is demanded to accomplish perfect decomposition of the bones; but it is important to guard against the possibility of any free sulphuric acid in the mass.

Another most excellent method of preparing bones for field use, is to dissolve or saponify the gelatinous portion by the employment of caustic alkalies. For this purpose, take one hundred pounds, beaten into as small fragments as possible, pack them in a tight cask or box with one hundred pounds of good wood ashes. Mix with the ashes, before packing, twenty-five pounds of slaked lime, and twelve pounds of sal-soda, powdered fine. It will require about twenty gallons of water to saturate the mass, but more may be added from time to time to maintain moisture. In two or three weeks the bones will be broken down completely, and the whole may be turned out upon a floor, and mixed with two bushels of dry peat or good soil, and after drying it is fit for use.

This mixture, embracing nearly or quite all the great essentials of plant food, is one which in its application will afford most prompt and satisfactory results. Its production cannot be too highly recommended.

The employment of bones in their raw condition, after grinding, has not generally been attended with results entirely satisfactory. Notwithstanding the published recommendations and testimonials, the fact remains, that the general verdict is not in their favor. My experience in the employment of this form of fertilizing material has been considerable, having used many tons during the past four years. Chemical analysis of corn and wheat, taken in connection with that of bones, would seem to show that they do not contain a sufficiency of the nitrogenous element to render them specifically beneficial to those cereals. And I have found, in practical trials, that they often exert but indifferent influence upon corn and wheat, when used uncombined or in a raw condition. This is especially true of steamed bones, where a portion of the gelatine has been removed in the manipulating process. When specifically employed upon soils appropriated to corn or other grain crops, failures, either partial or complete, have been often experienced; but upon those designed for roots, or some varieties of vegetables, success is uniformly certain.

Bones are made up of an earthy tissue of fine cells, in which an organic substance — gelatine — is enclosed. The gelatine holds the nitrogen, and undergoes putrefactive change, when moistened with water, with access of air. Ground bones undergo no change when air and moisture are excluded, and without this the powder is no more fitted or adapted for plant food than pebblestones or powdered glass. The putrefactive fermentation is attended with a copious evolution of heat; new bodies are formed, and disintegration of the structure takes place. The earthy constituents are composed principally of phosphate of lime. The best specimens I have met with gave, approximatively, of animal material and water, thirty-five per cent.; phosphoric acid earths, forty-seven per cent.; carbonate of lime, silica, &c., eighteen per cent. A direct estimation of the nitrogen gave, in one thousand pounds, of bones, fifty pounds; of phosphoric acid, two hundred and forty pounds; of lime, three hundred and thirty pounds. Hence we find they afford about twenty per cent. of nitrogen in their fresh condition. The phosphoric acid, however, greatly preponderates. Of this they furnish a rich supply.

In carefully studying the causes of failure of bones, when applied to the production of the cereal grains, it is evident we cannot *always* attribute it to want of the nitrogenous principle, as, in addition to what it is capable of furnishing, other sources of supply often exist in the soil fully capable of meeting deficiencies.

In considering some general causes which operate to prevent full and legitimate good results following the application of bones to soils, we shall see that the method or form of employment may have much to do with such failures. Adverse influences may be due, first, to adulteration in the bone material; second, the want of proper preparation before applying to the soil; third, unfavorable seasons. The first is an evil of very great magnitude, and one which can and ought to be abated. Every dollar accumulated by the industrious farmer is usually earned by the sweat of the brow, and he ought, particularly, to be exempt from peculation and fraud. Pulverized oyster and clam-shells, mixed so largely with bone dust by some manufacturers, exhibit a form of dishonesty particularly reprehensible, and is a source of great loss and disappointment to the husbandman. It is pleasing to know that all mill men do not practise this fraud.

The want of proper preparation is a fruitful source of failure. Bone dust ought always to be composted or rotted before using. It should be layered with good muck or soil, and kept moist until thorough decomposition results, and then it is fitted for the field. A gill of dry bone powder, placed in the opening prepared for a hill of corn, and covered with moist earth, heats rapidly; and I have found that in forty-eight hours a thermometer, with the bulb buried in the mass, indicates a temperature of 112° Fahrenheit.

This temperature is fatal to the germination of seeds, and besides the formation of caustic ammonia by the putrefactive change of the gelatine, furnishes an agent, when in excess and direct contact, equally as destructive as heat. Hence we learn why corn and other grains sometimes not only fail to flourish under its influence, but are absolutely destroyed in the germ. This heating, decomposing process should be effected prior to placing it in contact with seeds. The peat or soil used in connection with it effectively absorbs all ammoniacal and gaseous products, and holds them firmly until abstracted by the fibres of the plant roots in search of aliment.

It is not deemed important to present details of observations and experiments with bones or other fertilizers. In fact, there is much that is strictly empirical in such statements; they are entangled with so many modifying and distracting circumstances that they possess but little value.

The experimental labors undertaken upon my own farm have led me to adopt certain general conclusions as respects the teachings of chemistry and methods of employment of special fertilizers, which will be stated before I close. I have employed in these experiments a great variety of substances, under all possible forms and conditions, and have had regard to hygrometric and thermometric influences.

The analysis of soils constituted a prominent part of the labor, and it was in this direction that I expected chemistry would furnish most important aids.

It was soon apparent that but imperfect guidance was to be afforded by these analyses, however carefully conducted. In fact, the very perfection of the results, the exhaustive nature of the processes, created confusion and doubt, inasmuch as they revealed the presence of elements amply sufficient to meet the wants of plants; and yet they would not flourish in those soils.

Chemical reagents make palpable that which vital processes cannot force from their hiding-places. Acids dissolve hard and refractory substances; the tender spongioles of plants can only seize and appropriate those which are already in a state of solution. Hence, chemical research may demonstrate the presence in any given soil of the different forms of food which they require; but if the experimenter authoritatively announces to the farmer that it is fertile and capable of bearing crops, he is in danger of incurring contempt and ridicule, as practical trial disproves his science and his statements. The elements of fertility must not only be present in a soil, but they must exist in an assimilable form. To determine the presence and amount of the useful substances is not enough; research must proceed farther, and declare the *condition* in which they exist. There is very great liability to be misled in analyses of this character, and chemistry has failed to afford much practical aid to husbandmen in this direction.

I have found that soils holding but very disproportionate quantities of those elements which a particular crop required, would nevertheless produce it in fair abundance. I have to confess to disappointment to false predictions of results in some special instances; and until the true explanation presented itself to my mind, the matter of chemical research in soil analysis was under a cloud.

What was the explanation? Why, simply this: the soil, although holding the substance sparsely, yet all of it was in an assimilable condition; and as there was enough to meet the wants of a single crop, it was sought out and appropriated.

If the same crop had been repeated the succeeding year, it would have been very nearly or quite a failure. So long as chemical analysis of soils is inadequate to inform us respecting the condition, or how much of the contained plant food is in a soluble state as required by vegetable organisms, it will be impossible to make any certain predictions regarding its immediate or remote productiveness. Analysis must not on this account be discarded as useless or unprofitable in its teaching, as by its aid a vast number of significant facts have been developed, and many positive principles educed. A soil found to contain none of the constitutents which plants require could with safety be pronounced barren; and if there was an utter deficiency of any one essential, like phosphoric acid, lime, or potash, it could with equal safety be declared incompetent to support a certain variety of vegetation. Analysis fails to determine the positive immediate fertility of a soil, as we cannot determine how much material is in an assimilable condition. Viewing the matter as I do, it is not often necessary to resort to this expensive mode of inquiry. As will be shown, we can fertilize understandingly by chemical aids which do not pertain to the department of analysis. Chemistry not only unfolds the precise nature of soils, but also, as we have seen, the substances and principles which enter into plant structures.

The relations between the two are such, we are certain, that the inorganic matter found in the latter must have existed in the former. If there were no interfering agencies beyond our guidance, the whole problem of vegetable growth would be apparently the simple one of demand and supply, and this we could control.

It is an axiom which admits of no dispute or contradiction, that all the plant consumes of a mineral character comes from the soil. Let us consider for a moment the character of some grains — *wheat*, for example.

If we make chemical examination of wheat, we find that what we are able to rub off from the kernels, after moistening, with a coarse towel, is made up of woody fibre, and differs but little from the dry straw of the plant. The next wrapper, which is a continuous one, contains the most important constituents of the seed, holding the phosphate salts, and the nitrogenous ingredients. Here is stored up the little atoms of phosphate of lime, magnesia, soda, and potassa, which the microscopic mouths of the root fibres have sucked from the soil in which it grew. The office of the plant has been one simply of transference; it has transferred from the soil the earthy particles, — lifted them from their low estate to the highest within its power to attain, — placed them in position to meet the requirements of men and animals. Now, can the plant grow, and the seeds mature, unless the soil contains these salts? It may grow, and even luxuriantly; but shrivelled and imperfect seeds, few in number, will occupy the little pockets in the head, where, under the nourishing influence of a properly adjusted soil, the grains would round out with that plumpness that causeth the husbandman to rejoice.

It follows, then, that phosphoric acid is needful for the proper development of wheat seeds, — and, moreover, as the gluten which holds the salts is rich in nitrogen, that element is essential to its growth. These truths are a part of those which chemistry reveals to us respecting the constitution of the wheat berry. New England soils are deficient in these elements. Lime and the phosphates were never stored up in them in abundance, and through the successive croppings carried on by our fathers, men and animals have absorbed into their bony frameworks the little which had accumulated during the ages. The inference which seems to follow from these considerations is, that we have only to supply soil deficiencies, sow our wheat, and, casting aside all doubt and anxiety, patiently await the abundant harvest.

And why should we not do this? Have we not solved all necessary problems? Have we not learned by analysis what food is wanted, and have we not furnished it? Have we not learned precisely the constitution of the vegetable structure and its seeds? Do we not understand the nature of its appetite, and how it must be fed? Certainly we do. Why, then, should we meet with failures? Because we cannot bring under control all the conditions of vegetable growth. We could better command success were there no uncontrollable influences to be taken into account. The chemist cannot order meteorological agencies. He finds in his examination of plants, that they contain an abundance of water, and he also learns that vast quantities are constantly being exhaled during growth; and still another most important fact stands out for recognition: the food he supplies must be soluble in water, and, by its agency, voyaged through the microscopic canals to its appropriate resting-place. Water, then, is needful for perfect development of plants and seeds. Heat also must be supplied. The clouds must let drop the rain, and solar rays supply the diffusive warmth, else the husbandman returns from his harvests in sorrow, and science fails to aid him. Let us not unjustly condemn its teachings because it is unable to control the caprices of the seasons.

It is seldom, however, that crops utterly fail from the withholding of heat and moisture. Our fields are lean because of starvation — because we do not supply, through the soil, the food which plants require.

Chemistry teaches, what had already been learned from observation and experience, that in feeding vegetable growths, the kind of aliment demanded differs in different organisms. There are certain great families of plants which have diversified appetites, and they must be gratified in their tastes, or they refuse to bring forth their like. We know what they require, and we obtain hints as regards the best method of supplying their wants.

It is safe to follow the guidance of chemistry in fertilizing trees and vines. Careful examination of the wood and fruit shows what substances they most largely consume.

They differ from grains and roots not so much in the food they require, as in circumstance of condition. They are placed in the soil to remain for a series of years, and the consumption of certain elements is to be gradual, but constant. Therefore it is better to supply generously the specific aliment they require, and trust to soil decomposition for those articles of which the structure needs but a trace.

Near two years ago, I prepared a grape border sufficiently large for thirty vines. It was arranged in strict accordance with the chemical structure of the vine and fruit. Lime, phosphoric acid, potash, predominate in these; therefore, to meet the first want, mortar from the walls of an old building was used; for the second, well-rotted bone dust; for the third, ashes. But little animal excrement was employed, decayed sods supplying the needed humus.

Entertaining the idea that it is better not to make a homogeneous mixture of border materials, they were arranged in very thin strata or layers; first of soil, then bone, then soil with sand, then ashes, soil and sand again, then lime. The layers constituted but a mere sprinkling, and due regard was had to requisite quantities of each.

This bed was not disturbed with the shovel after it was completed. Arranged in this way, it seemed reasonable to suppose that the roots would not be required to travel so far for food in the early stages of growth, and that, extending as the supply failed, they would meet with a constant supply of nourishment. A kind of vegetable instinct evidently controls the feeders to plants, and enough push out to secure each distinct element in exact proportion to its wants; and the less the distance they travel, the less the vital force consumed in urging onward the nutritious principle.

The growth of wood the first season was strong and vigorous, and that of the past summer so extraordinary, that I had the curiosity to collect the wood that pushed out and matured from single buds, and weigh it, and the amount was found to be one hundred and seven pounds. Analysis of a portion of the leading shoot from one of the vines, basing the estimates upon ten grammes,* the amount employed in the examination, gave, as the quantity of water held in association, sixty-one pounds; combustible matter, forty-four and a half; *ash*, one and a half pounds. The ash contained, of potassa, twenty-nine parts in the hundred; phosphate of lime, nineteen; carbonate, thirteen; soda, three; magnesia, four; and small quantities of iron, silex, &c. The parts are given in round numbers, as, for the purposes had in view, scientific accuracy of statement is unnecessary. The wood, therefore, cut away at the fall pruning, carried off nearly eight ounces of potash, more than five ounces of phosphate of lime, and of lime and carbonic acid nearly four ounces. The subtile chemical agencies at work in the soil to render soluble and digestible so large amounts of mineral salts, how difficult to comprehend! and how amazing the

---

* Belonging to the French system of weights — decimal system.

amount of mechanical force exhibited by the vines starting from tender buds, capable of sustaining at maturity more than sixty pounds of water, and keeping it in motion through the pores!

It is fairly to be inferred from the results of this experiment that the luxuriant and healthy growth was due to the generous supply of food precisely adapted to the wants of the vine, and that the teachings of chemistry, as relating to the structure and appetite of vines, are worthy of regard.

What course is obviously to be taken, when, from repeated croppings, the grape clusters fail to appear? Shall we tear up our vines, as do many, and entirely remove the contents of the border as waste matter, and, at much expense, form a new one? Drenching with farm-yard manure from year to year has failed to restore to full fruitfulness; and why? Because it does not furnish in sufficient abundance the *one*, or, perhaps, *two* substances which are imperatively demanded. Contemplate for a moment the very large amount of *potash* stored up in the vines and fruit, — greater even in the latter than in the former, — and is there not palpable significance in this fact which chemistry unfolds?

Unleached ashes applied in generous quantity to old grape-borders will usually restore them to full fruitfulness, and render removal unnecessary. If they falter after the application, add finely-ground bones, and the work is done. The other agents needed are usually present in the border in sufficient quantity to meet all requirements; and it is only necessary to add those which have been removed by absorption to effect complete restoration.

If we can be as safely guided by the teachings of

chemistry in the cultivation of the three great families of plants upon which agricultural industry is most exercised, the cereals, — leguminous plants and roots, — we shall be directed by a light which will lead us out of all errors and all difficulties.

A recent English writer, in speaking of the results of the labors of Professor Ville, of the imperial farm at Vincennes, in the exuberance of his enthusiasm, exclaims, "There is nothing extravagant in stating that light has replaced darkness, that order has succeeded to chaos, and that the phantom of sterility is laid." Without sharing in such positive convictions in regard to the labors of the French experimenter, it is impossible to doubt or question the importance of his investigations. Indeed, in their general character, they can hardly be regarded as very new or novel; but they strike a death-blow at one delusion, which, like a spectre, has haunted chemists in their teachings upon agricultural questions for many years.

This relates to the empirical, indiscriminate application of single fertilizing substances to soils without any definite object in view. Perhaps the term "running for luck" will express the idea. The teachings of Sir Humphrey Davy, Liebig, Johnstone, Way, and many others, it must be confessed, have led in this direction, and thus established the uncertainty which invests such experimenting. The literature of agriculture is almost corrupted by disquisitions upon, and recommendations of, various salts or substances, as being the long sought-for elixir vitæ, the great specific, which is to retrieve all lands from barrenness.

It is quite certain that no such specific exists.

The recommendations of salt, lime, iron, nitre, ammoniacal salts, and a dozen other specifics in our numerous excellent and useful agricultural papers, cannot be regarded as beneficial to agriculture. It is quite natural for a soil cultivator, when, in the course of a series of experiments, he hits upon an article peculiarly adapted to the condition of his soil, to desire to communicate to others a knowledge of what has been so beneficial to him. The motives are honorable and praiseworthy; but he may thereby lead a neighbor into trying an experiment which ends in utter failure; and not only this, may do harm, by creating prejudice against that which, under a change of circumstances, might prove exceedingly useful. It will be understood that these remarks are made against the empirical use of single fertilizing substances. If any one has time, and inclines to experiment for his amusement, no harm can result, provided it be understood that the field of knowledge cannot be greatly extended by such labors, and that no observed beneficial results are of much use, except to the experimenter.

Perhaps gypsum may form an exception to these remarks. Because of its peculiarly isolated character, and of the uncertainty of its mode of fertilization, it must continue to be employed empirically until it is better understood. Gypsum has been the great stumbling-block in the way of chemists, and the question of its method of imparting fructifying influences to plants is still an unsettled one. The facetious author of a popular book upon husbandry remarks, "There has as yet been found no law by which to govern its application. On one field it succeeds; on another, to all appearances precisely the same, it fails. At one time it would seem as if its efficacy depended upon showers following closely upon its application; in other seasons, showers lose their effect. In one locality, a few bushels to the acre work strange improvements, and in another, *fifty* bushels work no change whatever. Now it is a hill pasture that delights in it, and again it is an alluvial meadow."

Liebig, after having advanced a very decided hypothesis regarding its mode of action, has in his more recent work, "The Natural Laws of Husbandry," abandoned it, and stated that "the whole matter is still involved in doubt."

It may not be worth while to add another theory to the many already advanced; but I cannot well help saying that experiment and observation lead to the conclusion that neither to the salt itself, nor to the separated lime or acid, is its fertilizing influence wholly or uniformly due. Its effects are nitrogenous in some cases. It is capable of furnishing nitrogen to plants, through the agency of an ammoniacal salt, resulting from soil decomposition.

My attention was drawn to the salt, accidentally, by observing a strong smell of sulph-hydric acid in a mass at the door of a plaster-mill. This had been trodden upon constantly, and water and mud containing organic matter become solidly impacted with it. Upon examination of a heap in the mill, I found that masses lying against wet timbers evolved the same odor. This led to experiment; and it was proved that gypsum in the presence of organic matter is readily deprived of its oxygen, and converted into sulphide of calcium.

It was further proved that this salt is capable of absorbing ammonia from the air, and from decomposing vegetable matter, and being thereby changed into hydrosulphide of ammonium; and this again may be changed into carbonate of ammonia by absorption of carbonic acid from the air. These are some of the changes which sulphate of lime is proved capable of undergoing, But this is not the time or place to protract the discussion. It seems to me probable, that the different theorists may be partly right and partly wrong; in short, that the salt is capable of exerting specific influence in several ways, according to the conditions under which it is acted upon. It may furnish nitrogen, or lime, or sulphur, or it may act on some soils physically, and not chemically, by absorption of moisture. If these views are correct, they may account for the doubt and confusion under which the question rests.

In all experiments with gypsum which have passed under my observation, the lands or soils, upon which its best effects are observed, are hilly pastures, with a northern aspect and a moist, moss-covered soil. Mossy meadows are greatly improved under its use.

The theory adopted is, that there must be organic matter in a moist condition, with ready access of air, in order to carry out those changes which have been alluded to. But I do not speak authoritatively upon this point.

After what has been said regarding the employment of specific fertilizers, it is probable the reader will have anticipated the recommendation I have to make, and that is, always to *compost* or *compound* elements of nutrition designed for plants, until a system is established, which will enable us to use single substances understandingly.

Chemistry, in its application to agriculture, has certainly made advances, inasmuch as it is now capable of demonstrating the correctness of two important propositions: first, that each field has its own peculiar wants; second, that each plant has its own peculiar appetite. It has further established its claims to respect and confidence, by showing that, meteorological influences being favorable, we can supply requisite food, in the proper quantity and condition, to secure the largest crops, with a great degree of certainty.

The system of Professor Ville, already referred to, embraces this idea. He proposes the use of what he denominates a *perfect manure;* that is, one made up of nitrogen, phosphoric acid, lime, and potassa. This, when made up and applied in proper quantities, he shows is capable immediately of changing a barren, silicious soil into one of perfect fertility.

I am willing to accept these results in general as in accordance with my own experimental observations during the past three years; or rather I yield assent to the correctness of the principle of producing and applying *perfect manures.* It is noticeable that magnesia is omitted as an element in his manure, par excellence. As we have before stated, there is present in most soils, or there is constantly being formed by decomposition, the minor substances, like iron, manganese, chlorine, &c., sufficient for the wants of vegetable organisms; but magnesia cannot be classed with them, as a glance at the composition of some important grains will show.

The ash of wheat affords twelve per cent., or twelve ounces in one hundred; the straw more than three per cent.; barley, seven; oats, ten; rye, ten; corn, eight; turnips, two. These quantities are large, and in the case of wheat grain come next to lime, forming one eighth of the whole amount of ash.

In countries where magnesian limestone abounds, the supply may be fully afforded by the soil. In France, Germany, and England this is probably the case; but in New England we cannot form a perfect manure and overlook the magnesian salts. In all the treatises and statements respecting fertilizing agents made by our chemists and experimenters, we find scarcely any allusions to the importance of the magnesian element; and this is indeed a matter of surprise.

It probably arises from the practice of copying the results of European writers — not from the deductions of original and independent research. Our soils are not constituted like those of Europe, and in the application of fertilizing principles they require different treatment. A perfect manure, then, adapted to our soils, should contain nitrogen, phosphoric acid, lime, potassa, and magnesia.

For the cereals, excess of nitrogen is demanded; for leguminous plants, as peas, beans, &c., potassa; for roots or tubers, phosphates. *All* demand lime and magnesia, and these must be supplied in the perfect food made ready for the plant-children of our fields.

Three questions remain to be answered: First, how shall we properly prepare these elements of nutrition? Second, how shall we apply them? Third, where can we obtain them? Chemistry is fully capable of answering the first. Apply all substances to the soil in the finest state of comminution; bring everything into a condition resembling as nearly as possible the excrementitious products of animals, which is the true condition. The bone for phosphoric acid must be reduced to an impalpable powder, and this is not its best form; it is better to dissolve it in acids or caustic alkalies, whose teeth are sharper than burr-mills or any mechanical levigators.

The *potash* must be in combination with carbonic acid, or in the form of carbonate of potassa. This is the most easily assimilable form, but in the caustic condition, as in ashes, it is readily changed to carbonate by contact with air.

The *nitrogen* must be furnished through ammoniacal compounds, or nitric acid salts. Lime, in form of phosphate, hydrate, or carbonate, may be employed, and the sulphate of magnesia furnishes the magnesian element in the cheapest, and in a sufficiently eligible form. How shall we apply them? This can be understood with a full knowledge of what end is had in view, or what special want is to be supplied. There can be no success under the ordinary conditions in which our agricultural labors are performed, unless an intelligent system is adopted and pursued persistently from one year to another.

It is not necessary that farmers should be practical chemists, to be successful in the employment of fertilizing agencies. A few simple principles, furnished by chemistry, if well understood and earnestly adopted, will enable any one to appropriate to his benefit all the important facts unfolded by science in respect to manurial applications. In treating a worn-out soil, a combination of all the elements needed for the three great families of plants should be employed; and if wheat or corn is to be cultivated, fields so prepared will yield a maximum return the first year. The second year, add the proper quantity of that which these grains demand in largest abundance, or which they abstracted from the soil the first year. These will be the *nitrogen* and *phosphate of lime*. If roots have been cultivated, the *phosphates* alone will be needed; if some member of the pod-bearing or leguminous family, potash. The three varieties of plants may be followed in rotation, with success, when by experiment the plan is clearly understood.

The great, prominent idea is, to maintain in the soil all the elements that plants require, and in sufficient abundance. If a particular crop removes a specific agent, supply it. Barn-yard manure furnishes all; and yet the same intelligence is to be employed in its use, and the equilibrium of elements must be maintained between it and crops. We know what and how much corn requires; we know how much good manure is capable of furnishing. A fundamental point in good farming is, to secure every ounce of this possible. It is an absurd notion, however, to suppose we can artificially produce it, by ill-adjusted mixtures of turf, sods, chaff, and rubbish. We can easily form a huge and dark heap, but if the salts be absent, — which almost alone give value, — it is hardly worth the labor it costs.

But the supply of barn-yard manure is not and cannot be adequate to our wants, and this brings us to the third question. If it was not for the matter of *cost* or value in known agents, which must always be balanced against the value of products, I could more satisfactorily answer this important inquiry.

At present, bones furnish the cheapest, in fact the only supplies of phosphoric acid, — ashes of potassa — ammoniacal salts, or nitrate of soda of nitrogen. In this country, prices of each of these are not yet so great as to place them beyond profitable employment; but unless the price of farm products continues to advance in a direct ratio with the rise of the agents, the time will come when their use must be relinquished. Chemists are hard at work upon some problems of great moment to the agricultural interests. These relate to the isolation of those principles of fertility which are locked up in the stony framework of our globe. Here we have reasonable grounds for expectation and hope; millions of pounds of potassa are reposing in felspathic rocks, and it cannot be long before they will be forced by chemical agents to relinquish their rich hoards of alkali. In the apatite and phosphorite minerals which abound so extensively in New York and New Jersey, we have abundant supplies of phosphoric acid and lime, and to them must we look for future wants.

There is not a single vegetable in the field or wood that does not contain in the ash potash, in some form of combination, and not a plant can be found upon our globe from which the phosphates are absent; therefore we must have full supplies of these indispensable agents.

We live in an ocean of gaseous matter made up of oxygen and nitrogen; seventy-nine pounds of the latter is contained in each one hundred of the mixture. Ready at hand, then, is this element; but unfortunately most plants are incapable of absorbing it in its free condition. Experiments have been made in France, which give promise of a supply of the ammoniacal salts, the nitrogen of which is derived from the atmosphere direct. If these chemical labors prove successful, we can understand through what source supplies of nitrogen may be afforded. Lime and magnesia are abundant everywhere, and these complete the list of important substances needed to render our fields inexhaustibly fertile.

We can hardly doubt as regards the abundant resources of nature, or cherish a hesitating faith in respect to a future supply of all our wants in feeding the plant-children of our fields. Our mother earth holds within her bosom all the various materials needed for the preservation and well-being of her children. When the woodman's axe ruthlessly stripped her of her rich vestments of umbrageous forests, and thus awakened apprehensions as regards the supply of materials needed to furnish household warmth, we were directed to the outcroppings of black carbon in our immense coal-fields; and when the Nantucket and New Bedford whalemen returned to their wharves, with the alarming announcement of the partial or complete failure of the ocean harvests of oil, the little rivulets of petroleum which oozed from the rocks in Pennsylvania were sounded to their depths, and immediately the oil spouted up in such quantities as taxed all our energies to secure. Let us look forward, then, with confidence, and trust to the future, and feel assured that chemistry, which holds the key which has unlocked so many rich chambers in the storehouse of nature, will open others fully capable of supplying all the wants of the husbandman.

CHEMISTRY OF THE SEA.

WHILE standing by the shore of the sea, contemplating its solemn grandeur, and reflecting upon its mysteries, we are apt to overlook some of the interesting and wonderful facts connected with its chemical history and character. It is natural that what is palpable to the eye, and so well calculated to awaken sublime and poetic emotions, should overpower the desire to study the "hidden things" of God, as connected with the great deep. It would be difficult at the sea-side to obtain listeners to a lecture upon the chemistry of the sea; but I venture to assume that under the less busy and exciting circumstances of home, the topic will not prove devoid of interest.

That which usually first arrests the attention of visitors to the sea, is the bitter and saline character of the waters, and the inquiry is made, From whence arises this remarkable condition? It may be said in reply, that it is but an exaggeration of that of ordinary lakes, and rivers, and springs; the same materials exist in them, only, in most instances, in infinitesimal quantities. As the atmosphere is the grand reservoir into which all gaseous or vaporous bodies pass, so the sea is the vast receptacle into which all the soluble substances washed from the earth are deposited. All kinds of soluble matter, washed out by percolating rains, descend to the ocean, by the agency of brooks and rivers; and as there is no outlet, no streams running from it, to carry them away, and as in the process of evaporation they are left behind, these soluble salts and minerals have been accumulating for ages, until they form prominent constituents of the waters. All bodies of water on the globe, into which rivers flow, but from which there is no outlet, except by evaporation, must necessarily be salt lakes. The Great Salt Lake, in Utah, that of Aral, near the Caspian, and the Dead Sea, in Judea, are remarkable examples of this kind. The Utah basin is filled with a saturated solution of this substance. This excessive saline condition is probably due to the existence of large bodies of salt in close proximity, or somewhere within the reach of streams that flow into it. Chloride of sodium, or common salt, is one of the most abundant of all the soluble substances found upon our earth, and consequently it predominates in sea waters. But while it is the most abundant and perhaps the most useful, it is by no means the only valuable substance carried into the sea. In quantity, next after salt, come certain combinations of magnesia, next salts of lime, the carbonate held in solution by excess of carbonic acid, then small quantities of potash and oxide of iron, and lastly, a trace of a most remarkable elementary body — iodine.

It seems a trifling and unimportant matter, this trace of the latter substance in sea water. The quantity is so infinitesimally small as scarcely to be recognized by chemical tests even after condensation by evaporation. Prior to the year 1812 this element was unknown. It was not found in plants, or rocks, or earths, or springs, in quantities appreciable to the chemistry of the last century; and even now we only know that a few atoms exist in the little watercress, and a few other aquatic plants, and in some springs and rocks; but from none of these sources could, probably, a single ounce be obtained. By the solvent power of water the minute quantities found upon the earth are taken up and deposited in the sea; and the Creator, as if foreseeing that this substance would be required in the arts to be cultivated by man, has provided a way by which it may be secured and appropriated to his purposes.

But before dwelling more particularly upon iodine, let us return to a brief consideration of the uses in sea water of some of the other soluble constituents. Everything in nature certainly has some palpable use. It is no accident or casual circumstance that the sea contains large quantities of the lime and magnesia salts. What stupendous results flow from this soluble carbonate of lime! Without it where could shell-fish procure their coverings, or the coral polyps the material for their curious structures? The shell of the clam, the oyster, the snail, the lobster, etc., is composed almost wholly of carbonate of lime: from what source do the fish obtain their calcareous coverings? Young oysters in two or three years acquire a size suited to be used as an article of human food. The little gelatinous speck floating in the water at birth has through some channel obtained two or three ounces of solid stone armor in the short space of thirty or forty months. It had no power to chisel it from limestone cliffs, and they are not always found in the vicinity of calcareous deposits. It has absorbed or drawn it from the water in which it moves; no other source supplies it. How immense are the beds of shell-fish upon the shores of the ocean! What a vast concentration of the lime, once held in solution, is effected by these feeble creatures, ranked among the lowest in the order of animate creation!

But still more wonderful is the work of the coral polyps. The geologist and the navigator will readily appreciate the extent to which the surface of the globe has been altered and modified, both in ancient and modern times, by the silent labors of myriads of these creatures, all engaged in the production of calcareous matter. The whole peninsula of Florida has been manufactured out of sea water by the little polyps. We are indebted to them for our marble houses, tombstones, and mantel-pieces. Powers's Greek Slave, pronounced by admirers of statuary to be "instinct with life," was probably once so in an *actual* rather than poetical sense. The marble is made up of the relics of these animals; and if from comminution they are not apparent to the eye, the microscope will show them. It is probable that nearly if not quite all limestone rock, in whatever form it is found, is of animal origin, and produced from the waters of the sea.

We now understand how vast quantities of lime are removed from sea water by the agency of living organisms: it remains to notice the channels through which iodine is separated, and placed in our hands for use in medicine and the arts. Human industry and science could never separate this element from sea water in any considerable quantity, and the power denied to man has been bestowed upon a slimy, repulsive *weed.* It is fortunate for us that the deep-sea plants have had conferred upon them a strange appetite, and that the food they seek is in part the sparsely disseminated atoms of iodine. It is probable that this constituent of sea water is in some way connected with the well-being of submarine vegetation, and that it is indispensable to its growth.

Through what feeble agencies stupendous results are attained! The little polyps build reefs and islands; the sea-plants (which every wave tears from their rocky homes), with their millions of open mouths, suck from the surrounding waters and appropriate as food tons upon tons of substances, otherwise unobtainable, and without which one of the most beautiful and important arts could have no existence. Seaweed possesses the remarkable power of abstracting from water, iodine. Let us inquire by what process of chemical manipulation it is forced to disgorge its precious treasures.

All *deep-sea plants* are more or less rich in iodine; but the *Palmata digitata*, that leather-like and greasy weed, with long round stalk and wide branches, has it in greatest abundance. The Irish call it tangle or lieach, and it is found strewn along our shores in large quantities after storms. But even this holds but a very small quantity. Every ounce of iodine upon the shelves of the apothecary has required at least *four hundred pounds* of weeds in its production. About thirty tons of the wet plants give one ton of *kelp*, as the incinerated mass is called, and from this nine or ten pounds of iodine is obtained. This would seem to involve a prodigious amount of labor and expense, bringing a high price upon the products. But the price is exceedingly moderate, seldom ranging in the English market above three dollars per pound. It would never pay at such prices to manufacture if the weeds did not yield other valuable products, as potash and soda. Without stopping to consider in detail the production of these salts, it may be interesting to know that probably more than *four thousand tons* of potash and *two thousand* of soda were introduced into the English market the past year, through the burning of sea-plants upon the coasts of Scotland and Ireland. The entire products of iodine from all sources must reach nearly or quite five hundred thousand pounds. How great is the industrial value of that which seems the most repulsive and worthless of all the products of nature! To what science are we indebted for opening up this great source of wealth? The reader's reply may be anticipated, — Chemistry.

The first work in the process is to collect the plants; they are then spread upon the ground and dried. Raked together in heaps, they are placed in rude kilns, made of beach stones, and burned. The red mass of ashes is stirred until it cools into a hard cake, called kelp, and is then ready for market and the interesting manipulations of the chemists.

The chemist breaks up the kelp into small pieces, puts it into immense tanks, pours on water, and leaches, until everything soluble is secured. He then evaporates the ley, and removes the different salts in the order of their solubility. First, sulphate of potash begins to crystallize; and that is removed while hot: as the liquor cools, the chloride of potassium begins to appear in beautiful white crystals; and that is removed. The ley is again boiled, and soon the soda salts appear; and they are removed; and now comes the iodine. If we commenced with sixteen hundred gallons of ley, we have reduced it to one hundred by evaporation and removal of the soda and potash salts: this holds the iodine in the form of iodate of soda and potassa. We must now free the iodine by taking up the soda and potassa (which it holds in combination) with sulphuric acid; accordingly, we add until it is saturated, and then we remove the yellow liquid to a style for sublimation. By the addition of heat the iodine is volatilized, or rises in vapor, and distils over into earthen receptacles, where it is condensed, and the process ends.

How often at the sea-side do we notice the disgust with which visitors thrust aside the slimy weeds, left upon the beach by the receding tide! It is probable that most carry about with them the photograph of some dear friend which they regard as a precious keepsake; unconscious, indeed, are they of the connection which exists between the light picture carried in the bosom and the marine plants trodden beneath their feet — a connection so intimate, that without the latter the former would probably be unknown. Iodine and its combinations form the basis of the photographic art; and this still resting undisturbed in the vegetable organisms, the splendid experiments of Daguerre would have been miserable failures.

CHEMISTRY OF A BOWL OF MILK.

IT is presumed that but few of those, in city or country, who sit down to the evening meal, consisting mainly of a bowl of milk, know anything of the interesting chemical nature of the liquid they consume. It must be plain, however, to the most indifferent observer, that it contains hidden supplies of nutriment of no ordinary character, as the most striking results follow from its use as an article of food.

The infant, in the earliest stage of existence, appears almost too tender and fragile to be raised from its downy pillow. In a few months, however, it becomes strong and lusty, the osseous framework is firmly knit together, the muscles are hard and flexible, the teeth grow, the nails and the hair push out, and all the high functions of life move on most vigorously. From whence come all the materials which, under the influence of the chemical and vital forces, accomplish such astonishing results? The bones must have an abundance of *lime* and *phosphoric acid*, and so must the teeth; the blood must have *iron*, and *soda*, and *potassa;* the brain demands *phosphorus*, in order that the embryo mind may be developed; the muscles need the nitrogneous element, and fat the carbonaceous. In addition, large quantities of water are needed, to maintain in harmonious action the functions of life and growth. Now, through what channel can these numerous chemical and nutritive elements be supplied to the feeble infant? The colorless and almost tasteless liquid which we call *milk* supplies them all, and usually just in the right proportions. In this nutriment, which has been provided for the young of the human race, and of the higher classes of animals, we have the most perfect type of food in general that it is possible to afford. With tender care, provision has been made for this helpless condition in life, and it is furnished in a manner which confers upon both giver and recipient the most placid enjoyment and happiness.

It is easy for any one to comprehend how it is possible to supply the wants of the adult organism, through the variety of food which is accessible and employed. The bread, flesh, fish, fruit, and vegetables, and even water, so largely consumed, may be easily understood to furnish the complex and diversified chemical elements which the system requires; but how *milk*, so simple in its physical characteristics, can embrace the essential nutrient principles of all forms of food, is not so easily comprehended. Let us for a moment glance at the composition of milk. It contains, 1st, a rich, nitrogenized material, *caseine;* 2d, *fatty* principles; 3d, a peculiar *sugar;* 4th, various mineral *salts*, principally consisting of phosphate of soda, phosphate of lime, phosphate of iron, phosphate of magnesia; the potash, it is curious to observe, exists in the form of *chloride* of *potassium.* The substances are held

CHEMISTRY OF A BOWL OF MILK.

in suspension by water. In one hundred pints of milk there are, usually, about eighty-eight pints of water. It is a remarkable fact that the composition of the milk of carnivorous animals, as the lion and tiger, does not essentially differ from that of the herbivorous, the cow, goat, &c. According to Dumas, however, there is no sugar in the milk of carnivorous animals. Cows' milk and human milk differ in the characteristic and leading constituent, *caseine*. One pint, or a bowlful, of the former affords about *three* fourths of an ounce, while the latter gives only *one* fourth as much of this important substance; therefore, in substituting cows' milk for the other, in feeding infants, it should be diluted with nearly two parts of water. Caseine is identical in composition with the muscular substance, and with the albumen of the blood, and it exists in milk in a *soluble* state. How easy it is of digestion and assimilation! The feeble powers of the infant are fully equal to its appropriation. By a molecular change of the simplest kind it becomes the material of flesh, or passes into the cellular tissues by an act of oxidation. Hence come muscular strength and nervous energy to the young offspring. In childhood the function of respiration is exceedingly energetic, and ordinary food, in ordinary quantities, would be hardly equal to the waste. But in milk we have provision made for this demand. We have *two* non-nitrogenous bodies, *butter* and *sugar;* these *burn*, in the body, to carbonic acid and water, and develop the necessary heat. In one hundred ounces of milk there is about half an ounce of mineral salts. More of the lime being needed to form the body structure, it is furnished in milk in large excess of the other salts, so that the growth of the bones keeps pace with that of other portions of the body. The trace of ferruginous matter is all that is needed to supply the blood with the little iron ships, whose offices are to load with oxygen in the lungs, and voyage it through the great ducts to the capillaries, where the butter and sugar are oxidized or burned for warmth. The phosphate of soda and the chloride of potassium find their appropriate place in the blood and secretions, and perfect harmony and efficiency in chemical and vital changes are secured. Nothing superfluous is to be found in milk, and nothing essential to the well-being of the infant has been omitted.

What is man, or an animal, but a kind of chemical laboratory, where transmutations and changes in gross matter are going on constantly, in order that force may be developed, and the machine or body kept in motion? Is an atom of iron, or potash, or soda, any more sacred, or entitled to higher consideration, because it has happened to be absorbed from the rocks or dust by vegetable growths and taken into the body, there to be manipulated by the unseen chemist, and perhaps assigned, for a brief period, a place among the other earthy or atmospheric constituents of the flesh? What is *health* but an undisturbed play of chemical affinities in the animal organism? What is *disease* but imperfect chemical reactions, or insufficient supply of necessary chemical agents in the same?

With this brief and imperfect view of the chemistry of milk as an article of food, let us for a moment look at some of the physical and chemical changes it is capable of undergoing in the various processes to which it is often subjected.

Caseine is a very remarkable substance, and is found only in milk, where it exists in a state of perfect solution. It is held thus by the presence of a small quantity of alkali. Now, if we add to milk a few drops of acid, we neutralize this, and the caseine coagulates or forms a solid body, which is called *curd*. The manufacture of cheese depends upon this coagulation of caseine. This result, produced under the influence of a *simple wet membrane* without acids, is a phenomenon so remarkable that it is no wonder it has excited much attention. A bit of the lining of a calf's stomach, — *rennet*, — placed in milk, precipitates the caseine rapidly, and from this cheese is formed.

Berzelius states that he took a small piece of this membrane, washed it clean, dried it as completely as possible, weighed it carefully, put it into eighteen hundred times its weight of milk, and heated the whole to 120° Fahrenheit. After some little time coagulation was complete. He then removed the membrane, washed, dried, and once more weighed it; the loss amounted to rather more than one seventeenth of the whole. According to this experiment, *one* part of the active matter dissolved from the membrane had coagulated about *thirty thousand* of the milk. Does chemistry explain satisfactorily this wonderful effect of infinitesimal quantities of rennet upon milk? It does. The change is due to the presence of "sugar of milk" in the milk. This substance is peculiarly prone to pass over into lactic acid, under favorable conditions, by appropriating the elements of water. The membrane acts as a *ferment*, lactic acid fermentation is set up, and a minute quantity of that acid is produced; this immediately acts upon the caseine, coagulating it and producing curd. Without the aid of the membrane milk will precipitate the curd. There is no lactic acid in fresh milk, but, after a few hours in a warm place, it makes its appearance, the caseine falls, and it becomes *sour*. This could not occur if no sugar was present in the milk. The thin, pale-colored, translucent liquid remaining after the curd is removed, called "whey," consists mainly of water, holding the saline constituents and the sugar of milk. The curd, after it is salted and pressed, undergoes a particular kind of putrefactive change, which gives flavor to the cheese.

Milk, examined by the aid of a microscope, presents to the eye myriads of remarkably minute globular particles, suspended in a thin liquid. These particles are termed *butter*, and rise to the top upon standing, bringing with them a portion of the caseine and serum, and thus form *cream*. By agitation, or churning, the fatty matter is separated from the milk, and butter is produced.

The secretion or production of milk may be very seriously and detrimentally interfered with. By the employment of certain articles in the food, the *color*, *odor*, *taste*, and *medicinal effect* of milk may be modified; and this is so well understood by physicians, that in France children are brought under the influence of medicine administered to the mother. And further, a new form of treatment has been instituted, which is based upon the plan of administering to animals certain remedial agents, and causing patients to live upon the milk of the animals. It is evident we cannot be too strongly impressed with the importance of providing pure, healthy milk for children. The state of health of the female has much to do with the quality of the milk; and a sickly mother should hesitate before jeopardizing the well-being of the infant by allowing it to feed at the maternal fountain.

It is equally as important that cows' milk should come from perfectly healthy animals. Labillardièro states that the milk of a cow, affected by a species of phthisis, contained *seven times* more phosphate of lime than usual; and Dupuy also noticed the large quantity of calcareous matter in milk from cows similarly affected. Diseased milk may be known by its want of homogeneousness, an imperfect liquidity, a tendency to become viscid on the addition of ammonia, and, on microscopic examination, the presence of certain globules not found in healthy milk.

The adulteration of milk, by additions of water, is a very common practice by milk-venders in cities. It is a matter of regret that, owing to the great inequality in the amount of water found in cows' milk, the conviction of offenders in court is rendered a matter of so much difficulty. Much, however, may be done, by vigilant, well-directed efforts, to arrest the monstrous frauds in milk in our cities.

CHEMISTRY OF THE DWELLING.

THE chemistry of the dwelling is a subject which should interest every person living in a civilized country, and surrounded with the household blessings and comforts which science and art confer. And yet how few there are, comparatively, who have studied or inquired respecting the chemical processes going forward, and the devices and appliances of modern science, which contribute so directly to our well-being and comfort within the walls of our own dwellings. Let us, while perchance the storm howls fitfully without, draw closer around the parlor fire, and consider, for a little time, the wonderful and beautiful chemical processes which we witness upon either hand, and which, if suspended for a single day, would be productive of so much discomfort and danger. Without the intense diffusive light proceeding from the burning of the oil, or wax, or spirituous liquid upon the table, or of the invisible gas from the suspended jet, we should be unable to gaze upon each other's happy faces, or read the pages of a book, or pursue, after nightfall, the usual avocations of the family. Without the blazing coals within the grate, or the wood upon the hearth, or the warm air or steam passing into the room through the proper channels, we should become chilled and benumbed with cold, and disease and death would supervene.

We are all constant chemical experimenters, although we may be unconscious of the fact. The lecturer, surrounded with his strange compounds and curious apparatus, delights us with his attractive and brilliant experiments; and yet, there are but few more interesting or wonderful than we perform in lighting our parlor fires, preceded as the act usually is, by the ignition of a common match.

Fifty years ago the lighting of a match by the slight friction necessary, would have been regarded with amazement, and any public exhibitor of the experiment might have been punished as a necromancer. There are thousands living, whose knuckles have been torn with the old flint and steel, who remember the progressive introduction of quicker and better methods of producing fire. Our grandfathers, when the tinder in the horn was damp and obstinately determined not to "catch," were accustomed to take down the old "King's-arm" from its dusty resting-place upon the wall, and flash gunpowder by the aid of its massive, rusty lock, thereby procuring fire to warm their breakfast of porridge or johnny-cake. It was no unusual occurrence to hear a report, and see the big powder-horn fly up chimney, simultaneously with the click of the lock, as our grand-parents, with all their virtues, were *careless*, like other men.

In this advanced age (thanks to chemical science), we have a more excellent way; and now let us, as a matter pertaining to the chemistry of the dwelling, describe briefly the history and chemistry of the friction match. The invention of the phosphorus match was preceded by others less convenient and more uncertain in their character, but all a vast improvement upon the flint and steel. The history of the match forms no exception to the rule, that all discoveries are progressive in their nature, that all products of the inventive faculty must pass through the chrysalis state before reaching entire perfection.

While people in affluent circumstances in cities were indulging in the use of the *fire syringe* and the *acid bottle*, to produce fire, contrivances which were regarded as marvels in science, there appeared, about thirty years ago, in the market, a little square paper box, containing two dozen strips of wood with a mass of black, ugly-looking composition upon the end of each.

A piece of sand-paper was found carefully folded in the top of the box. One of these matches, drawn rapidly through the sand-paper, ignited with a slight report. The price per box, upon their first appearance, was one shilling, and the manufacturers were busy for a time in supplying them at this exorbitant price. The sensation created by their appearance was about equal to that produced by Franklin, at the time of his discovery of the electricity of the clouds, nearly a century since. These were the famous Lucifer Matches, the worthy predecessors of the friction match of the present time. Perhaps it would have been better that invention had gone no further: certainly, this method of producing combustion was rapid and easy enough; but some considerable pressure was required to produce the necessary friction, and sometimes the top was pulled off without being ignited, the sulphurous antimonial vapor was regarded as pernicious to persons with weak lungs, and so, upon the appearance of its great rival, in a few months it fell into entire neglect.

In obtaining fire, or causing combustion, the most ready method is by the use of friction. The spark following the stroke of the flint upon the steel is produced by friction. A minute portion of the steel is clipped off by contact of the flint, and it is rendered incandescent, or heated to a white heat, by the concussion; this falling upon tinder, or a thin film of carbon, it is set on fire.

Friction raises the temperature of bodies, and some bodies burn at so low temperatures, that the slightest movement across a rough surface is all that is requisite to cause them to burst into a flame. Sulphur and phosphorus are bodies that inflame at low temperatures, and these are consequently employed in the manufacture of matches.

The latter is a most remarkable element; its greediness for oxygen is so great that it attracts it, and burns spontaneously in contact with air. In the manufacture of a phosphorus match, a splint of light wood is dipped in melted sulphur, after drying, it is again dipped in softened phosphorus. If left in this condition, it would be vastly more dangerous than in its finished state, and would be entirely unsafe to harbor in our dwellings. To prevent spontaneous combustion and protect the phosphorus from contact with air, the match is again dipped in gelatine or glue, which is the third and last coating it receives. In igniting it, the friction disrupts the film of glue, raises the temperature of the phosphorus so that it burns — this in turn ignites the sulphur, that the wood, and thus the beautiful experiment of producing instantaneous flame is complete.

The discovery of phosphorus, and of easy and rapid methods for its manufacture or isolation from the bones of animals, is among the most striking and important of the triumphs of chemical science. How apparent the wisdom and goodness of the Creator in calling into existence an element of such singular properties, so inflammable that the warmth of the hand is sufficient to cause it to burst into flame! As if fearful that so dangerous as well as useful a substance might prove an enemy rather than a friend to the race, before progress had been made in the arts of civilized life, He diffused it very sparsely in the ancient rocks in such condition as to be entirely unobtainable without the aid of science. The conditions upon which we, with all our skill, are enabled to procure it in quantities, are peculiar. We do not go to the rocks for it, but are compelled to wait until, by the operations of nature, it is dislodged from them, and fitted for plant aliment or food. Transferred from ancient lavas, and plutonic masses, to plants, it is consumed by animals, and, passing through the circulation, it finds a resting-place in the bones, from which, by calcination and other processes, the chemist obtains it in large quantities. How circuitous is the path it travels before it finds lodgment upon the end of a match! Truly, we cannot but regard a thing so common with interest, when we remember its origin and chemical history.

In the ignition of the match we have set in motion a series of changes which have resulted in the *burning* or destruction of its substance. In its employment to ignite the combustibles in the grate, or to light the jet cf gas, or the candle, or lamp, we have thereby caused an activity of slumbering chemical forces, which are slowly producing in those bodies similar changes. We notice that they continually waste away — we see the ashes cleave off from the mass of coals or wood, the falling of the line of oil in the lamp, and the candle's flame burns lower and lower until it reaches its socket — and expires. These changes we notice, and, untaught by science, should be left to suppose that we had destroyed or annihilated a small portion of the materials of God's universe. Science teaches that this is entirely beyond our power. However strong and mighty man may be in modifying or controlling the elements created by Omnipotence, he can never create or destroy a single atom.

Since that auspicious moment when sunlight burst through the chaotic darkness which enveloped our planet, we have reason to believe that nothing has been added to or taken from its mass.

If our sense of sight was competent to observe the invisible operations of nature, we should see in our parlor fires not only the flame, and the smoke, and the ashes, but those subtle exhalations, the products of the burning, which pass up the chimney and become dissipated in the ocean of air without.

How apparently desirable would be an acuteness of the visual organs, so that we could see the little infinitesimal atoms of matter grouped and compacted together, forming coal, and oil, and wax, and tallow, all ready for the warm embrace of the oxygen of the air, which, by uniting, rends them ruthlessly asunder. Seated in our parlors, we could watch at our ease the elemental changes, the dissolution, and the new birth of bodies during the process of combustion.

From an enumeration of the different kinds of little atoms or elements in the wood, coal, gas, oil, tallow, and wax, we learn that out of the sixty, which the Creator employed in constructing the world and all things therein, he has made use mainly of but *three* in forming these substances. Therefore, in all that concerns the chemistry of light and warmth, we have to study the changes and modifications of but three different kinds of materials — a field apparently circumscribed and easily explored.

The gas, oil, tallow, &c., differ in composition from wood and coal in being formed from carbon and hydrogen only, without any ash-forming elements. These two are grouped together in slightly varying proportions, by the burning of which we obtain light and heat. The burning of these two elements is produced by a third body rushing in whenever the temperature is raised to a certain point, and, violently uniting itself with them, producing, by the union, extraordinary warmth; which, diffusing itself, is very agreeable to our cold, benumbed bodies in winter.

This third powerful combatant or element is *oxygen*, the most important and essential of all the material creations of the great Architect. It is gaseous in its nature, and, although unseen by human eyes, it plays a most conspicuous part in the great operations of Nature. Its importance may be understood by a contemplation of the fact, that, in connection with nitrogen, it forms the vast volume of the atmosphere, and, in combination with hydrogen, water, which, in the shape of oceans of unknown depths, lakes, and rivers, occupies three fourths of the surface of our planet. The solid earth we dwell upon is chiefly made up of oxygen, in union with silicon, aluminum, and calcium, the metallic base of lime.

The carbon and hydrogen materials burned in the parlor are seized and consumed by the oxygen ever present in the air. It exists there in a free, aeriform condition, and seems to be ever upon the watch for heat-producing agencies, so that it may be enabled to fix its corrosive teeth upon the wood or coal to rend them asunder.

We wish to observe the changes attendant upon the process of burning, from the beginning to the end. We have learned from experience that wood takes fire and burns more readily than coal, and chemistry affords the reason. It is because it contains a greater number of inflammable atoms of hydrogen than coal, and the softer the wood the easier it ignites; hence we place splints of wood or shavings at the bottom of the grate, and upon this the coal. The hot hydrogen flame of the wood soon ignites the hard carbon of the coal, and the whole is in an active state of combustion. The invisible oxygen around the pile rushes in, drawn by an irresistible affinity; the infinitesimal atoms of hydrogen yield first to his embrace, then the carbon, and both by the union are instantly metamorphosed or changed into new and very different bodies.

These new bodies that are produced are called the *products of combustion*. In this hot contest for new combinations, we notice that it requires eight atoms of the oxygen to master one of the hydrogen; seven, six, or five cannot appropriate the single hydrogen atom to form the desired union; the eight join themselves to the one, and the nine, with the rapidity of the lightning flash, are changed into an atom of *water*.

The carbon of the coal and wood unites with the oxygen of the air in two proportions. When atom joins atom, a new substance, called *carbonic oxide*, is produced; but when two atoms of oxygen join one of carbon, a sour, poisonous body results, called *carbonic acid*. These three *new* bodies, the water, the carbonic oxide, and carbonic acid, become heated and ascend the chimney-flue and diffuse themselves through the air without.

In the escape of these bodies we lose no substance that can be more completely burned or changed, with the exception of the carbonic oxide. This gaseous body can and should be burned again before it is allowed to escape, for it is capable of affording us a further supply of heat. If we can compel the carbon, by increasing the heat, and opposing barriers to its escape, to take up two atoms of oxygen instead of one, we thereby burn or oxidize it completely, and we obtain all the heat possible. If the wood or coal were burned *completely*, there would be but *two* products of combustion besides the ashes, viz., water and carbonic acid. This important matter of the proper method of burning coal will be alluded to again.

The ashes we observe cleaving off from the fuel contains the earthy matter which the tree obtained from the earth during growth. Besides silex, or sand, the ashes of wood contains potash, so important in the manufacture of soap. In coal there is usually found a trace of sulphur, which, in burning, unites with oxygen, forming sulphurous acid.

Let us look at the coal and the wood upon the hearth with our vision quickened and perfected, so that not only will the composition of these substances be apparent, but the whole process of burning also. In the coal we observe black *carbon* predominating in the aggregation of atoms. In the hard anthracite there are ninety-one little atoms of carbon in the hundred to nine of other elements. The nine are seen to be hydrogen, and those that form the ashes. We notice a difference in a soft bituminous coal, there being a less number of carbon atoms and more hydrogen.

Wood, although entirely different in color, is seen to be made up of the same elements. Hard woods, like oak and hickory, have the larger, while soft pine and maple have the smaller number of atoms of carbon. All varieties of wood, however, have a much larger amount of inflammable hydrogen than the hard varieties of coal.

A beautiful experiment, indeed, do we perform in instituting the process of combustion. Interesting and attractive as it is in itself, how seldom should we indulge in it, did not the necessities of our existence demand its constant repetition! The production of *heat* is the great end we have in view in kindling and maintaining our parlor fires. Man in a savage state in tropical climates is quite independent of the uses of fire; but such is not the case with civilized man. He finds it necessary in the preparation of his food, although the body is heated by the intense solar rays of the tropics.

Heat is a constant attendant upon the burning process. Whenever and wherever the element oxygen joins itself to hydrogen or carbon, or any of the inorganic elements, as iron or zinc, the union is attended by the evolution of heat. These elements will always rush together, and burn and destroy each other when circumstances permit. There is a class of compounds denominated oxides or rust. Oxygen, uniting with iron, forms rust of iron. In the union of the one with the other, a fixed amount of heat is evolved, no matter whether the rusting process goes on slowly or rapidly. The iron implements and vessels found at Pompeii, which have been slowly burning or rusting for eighteen centuries, have evolved as much heat in the process as would have resulted, had they been burned instantaneously in an atmosphere of pure oxygen. The process of decay in vegetable substances is an oxidizing process, and heat is evolved. Water, properly speaking, is the rust of hydrogen, and in its formation, by uniting with oxygen, an enormous amount of heat is developed. The union of the oxygen of the air with the coal in the grate, or the wood upon the hearth, produces the same phenomenon. It would be gratifying to know more of the nature of heat, and also of light and electricity; but, since it is denied us, we may indeed be grateful that the beautiful principles and changes involved in combustion are so clearly unfolded by science.

The fallacies of past ages, as it respects correct knowledge of natural phenomena, are in no way more forcibly illustrated than in the prevalent theories respecting combustion. Before the discovery of oxygen gas, it was explained by supposing that all bodies contained a principle called *phlogiston*, the presence of which enabled them to burn. When a body burned, it was supposed phlogiston was liberated, and that when it lost phlogiston, it ceased to be combustible; it was then said to be *dephlogisticated*. The heat and light which accompany combustion were attributed to the rapidity with which the principle was evolved.

If such an hypothesis were correct, the coal or other burned body ought to weigh less after the process; whereas it was found that the results of combustion were *heavier* than before the combustion took place. As soon as methods were devised by which the three great classes of acids, alkalies, and oxides, the products of combustion, could be secured and examined, the theory was disproved.

There is no known method by which heat can be measured. The heat evolved depends not upon the coal or wood, but upon the quantity of oxygen which enters into combination with them in burning. The oxygen supplied to the coal or wood, is obtained from the room or apartment in which the burning process is going forward. This, of course, would soon fail to furnish the requisite amount to the fire, were there no sources of supply. Through some avenue, the air from without must find its way into the parlor to feed the fire, and furnish oxygen to the lungs of the occupants. The usual places of ingress are the cracks and crevices of the doors and windows. A window, with the usual accuracy of fittings, will allow about eight cubic feet of air to pass into the room each minute; an ordinary door will admit rather more if it open directly into the air. When we reflect that each individual in a room ought to have at least four cubic feet of air per minute for respiration, and that

CHEMISTRY OF THE DWELLING. 75

during the evening every source of flame as large as one candle vitiates one cubic foot more, we see how important a good supply of air is for other purposes than to afford oxygen to the fire. I would not wish to be understood to say that each individual's respiration converts the oxygen of four cubic feet of air into carbonic acid each minute, but that that amount is rendered unfit for further respiratory use; every pound of hard anthracite coal burning in the grate absorbs from the air of the room about two and one half pounds of oxygen, and at least fifteen pounds of air is deoxygenized to furnish it. A parlor of common size, twenty feet by thirteen, and ten feet high, contains about two hundred pounds of air; it is evident, since three and one half pounds of carbonic acid are produced from each pound of coal, that, if it were permitted to permeate the room, it would render one fourth part of the air, at least, poisonous. This, diffused throughout, would cause death to the inmates in a short period of time.

How impresssive the fact, that by the marvellous chemical processes which we are obliged to institute to render our climate habitable, we call into existence an agency which is potent to destroy almost instantaneously! If, upon a cold winter's day, we consume, in our parlor grates or stoves, twenty pounds of coal, there has been poured out upon the air seventy pounds of poisonous carbonic acid, which would render irrespirable, if not diffused beyond a point contaminating and destructive, about two hundred and eighty pounds of air. When we contemplate this fact, and reflect upon the thousands and tens of thousands of open ducts, which are pouring out the poisonous exhalation in enormous quantities in large cities, a momentary feeling of apprehension pervades the mind, and the knowledge that the specific gravity or weight of the deadly gas is greater than the air, does not diminish that apprehension. But chemistry dissipates our fears, and points to the wonderful provisions of the Divine Author to avert the apparent evil.

If the heavy carbonic acid so copiously evolved, were simply to obey the natural laws of gravitation, and descend into the streets of cities and large towns, a most dreadful asphyxia would instantly seize upon every man, woman, and child, and in the short space of a few moments, not a breathing inhabitant would remain. But the law of gaseous diffusion comes in here, and shows us that there is a " higher law" than that of gravitation, which is intended for our preservation. By its irresistible agency, the heavy poisonous gas is not permitted to fall, but, at the moment of its production, it is blended and diffused through the mass of air, upwards as well as downwards, and is wafted by the winds in all directions. The wonderful nature of this law of gaseous diffusion is forcibly illustrated by experiment. If we take two gases of most opposite qualities, as it respects weight, carbonic acid gas, and hydrogen, and place them in two vessels communicating with each other by a narrow tube, we shall find in a very short time that perfect mixture has occurred. This will take place if we reverse the order of their specific gravities, by placing the hydrogen in a higher vessel, and the carbonic acid in the lower; a wet membrane may divide them, and we shall prove that there is a strange tendency to unite. Carbonic acid is more than twenty times heavier than hydrogen, and it

CHEMISTRY OF THE DWELLING. 77

would seem that while the tendency of the former must be downwards, the latter would be upwards.. But such is not the case; they shortly become thoroughly blended together. This law holds good in the mingling of all gases of different densities which have no chemical action on each other.

Thus is carbonic acid equally diffused through the whole atmosphere. There is enough constantly present in all parts of it to form a stratum or bed, thirteen feet thick, over the entire earth, should it descend and occupy that position. It is not necessary for the mind to revert to fire, as an agency more potent than others, for the destruction of the race. The terrible nature of such a layer of heavy irrespirable gas was most forcibly illustrated to the mind of the author by an examination of the great vats filled with it to the brim in the immense brewery establishment of Messrs. Barclay & Co., in London. A plank was displaced by the attendant, and it was allowed to flow over the side, like water over a fall; and a single inspiration produced vertigo, and other unpleasant consequences. At the celebrated *Grotta del Cane*, near Naples, a dog is often thrown into a cave filled with the gas. After a few painful respirations the poor animal is apparently devoid of life, but subsequently recovers, upon being dragged to the pure air without. The experiment is a cruel one.

Whilst the carbonic acid, resulting from the chemical changes of the coal and wood of our parlor fires, is so fatal to man, it is absolutely indispensable to the existence of vegetable life; and if it were withdrawn from the atmosphere by being absorbed by water, or in any other unusual way, the latter must cease from the earth altogether, and with it all animal life. Plants depend in a great measure for their sustenance upon the atmosphere; and the carbon of the wood and coal which we have watched through the changes attendant upon combustion, and which resulted in the production of a poison, is obtained by the plants from the air by decomposing the same deleterious body.

Paradoxical as it may seem, it is quite evident that we are dependent for *life* upon the very poison which is so pregnant with *death*. The production of carbonic acid from the various sources upon our planet, is so marvellously balanced by the demand for the same for plant aliment, that there is no perceptible change in amount in the air from year to year, there being uniformly about one two-thousandth part by measure present.

It is evident, from the foregoing facts, that while the products of fire may be safely discharged into the air, it would be productive of the most fatal results to allow them to escape into the rooms of our dwellings, to be breathed into the organs of respiration. The ingenuity of man has devised an arrangement of flues which subserve the double purpose of conveying away the noxious gases and creating an upward current of air, which, passing through the ignited materials, draw oxygen towards them, and increase the intensity of the flame.

Simple as is the contrivance of a chimney, it is singular they should be a modern invention. There is no record of any chimney being used in dwellings prior to the twelfth century, and even as late as the time of Queen Elizabeth they were quite uncommon in England. It is stated that Good Queen Bess herself resided in a room unprovided with the luxury of a chimney. They were undoubtedly in use in Venice in the middle of the thirteenth century, and in Padua, but not in Rome; for when, in 1368, Cararo, lord of the first-named city, visited Rome, he found no chimneys in the inn where he lodged, and his host kindled a fire in a hole in the middle of the floor for his comfort, or rather discomfort. The buried cities of Italy afford no evidence that chimneys were used by the ancient Romans, as no contrivance has yet been discovered in either Pompeii or Herculaneum designed to carry away the products of combustion. Before the construction of chimneys, the smoke was allowed to escape through an orifice in the side or top of the room. And in the imperfectly constructed dwellings of those times, there were plenty of vents for the ingress of air, so that smoke and gases were diluted, and rendered comparatively innocuous.

We may almost presume that smoke was a *luxury* in those early days; the people certainly regarded a smoke-impregnated atmosphere as a healthful one. Old Hollingshed, an Englishman, who wrote several centuries since, thus complains of the innovation of chimneys: —

"Now we have many chimnies, yet our tenderlings do complain of rheums and catarrh, and poses. Once we had nought but a rere-dose [a fireplace], and our heads did never ake, for the smoke of those days was a good hardening for the house, and a far better medicine to keep the good man and his family from the quack or pose, with which then very few were acquainted. There are old men yet dwelling in the village where I remain, who have noted how the multitude of chimneys do increase, whereas in their young days, there were not above two or three, if so many, in some uplandish towns of the realm. And peradventure in the manor places of some great lordes, but each one made his fire against a rere-dose, in the hall where he dined and dressed his meat.

"But when our houses were built of willow, then we had oaken men; but now our houses are built of oak, our men are not only become willow, but a great many altogether men of straw, which is a sore alteration."

The quaint, humorous old writer would be called a "croaker" in these days. He was evidently one of those who believed in the rapid deterioration of the race, and was disposed to charge it to the effeminacy of the times, by which many were led to refuse to breathe an atmosphere saturated with smoke and cinders, — a philosophy worthy of the fourteenth century. Whilst it was possible to dispense with chimneys, so long as wood alone formed the only combustible material, the introduction of coal at once rendered them indispensable. The large quantity of volatile sulphuretted gases which are formed by the heat, and which pass off from soft coals, together with the carbonic acid gas proceeding from all varieties, would render rooms positively uninhabitable were no chimneys in use. The visible smoke proceeding from burning wood, composed as it is mostly of fine cinders and unchanged particles of the wood, is not poisonous, but in a very considerable degree irritating to the mucous membrane of the air-passages of the mouth and nose, and also to the eyes.

Hence those living in smoky houses have the impression that they are troubled with continued catarrh or colds, the irritability produced by smoke resembling so closely that resulting from this affection. Notwithstanding the statements of Hollingshed, our experience leads us to believe that the lungs are rendered more sensitive to atmospheric changes by the frequent inhalation of smoke; and those compelled to live in a smoky atmosphere are more troubled with "rheums and catarrh, and poses" than those who do not.

One thing is certain — no annoyance is regarded as more severe than a smoky house; and if the ancient philosophy of the "hardening" process was correct, few would submit to it for the benefits conferred. How to make a chimney draw well is a question of the first importance with thousands, and one to which the sagacious Franklin early directed his attention. He was regarded in England at one time as the most accomplished "smoke doctor" living, and his advice was sought upon the subject of draught in chimneys with great frequency.

A few simple principles are worth remembering respecting the *cause* of draught and methods of increasing it. If a chimney is constructed of any height and dimensions, it is of course filled with air. And if the column of air within it weighs as much as a column of equal height surrounding it without, it will have no draught. Two things operate to change the relation of the columns, and create an ascensional current within the chimney. One is elevation or height; the other, warming the air by fire, by which it becomes rarefied, and its weight diminished. The taller the chimney, or the hotter the fire, the more rapid will be the draught. It must be constructed vertically, as much length horizontally, by cooling the air before it gets into the effective part of the flue, will be sure to spoil the draught.

If a grate or fireplace is troublesome by reason of incompetency to convey away smoke, it may be owing to too great an aperture above the fire, so that a large volume of cold air enters the flue without passing through it, and thus is constantly cooled. A stove connecting with it would work satisfactorily, because the air would be compelled to pass through the fire, and thus keep the chimney current warm and active. A sliding valve, so arranged as to increase or diminish this orifice *above* the fire, is often a complete remedy.

Chimneys upon the north part of a building do not uniformly work as well as others, because of the refrigerating influences of the locality. A chimney thus situated may be made successful by constructing it double, or making an air chamber around it to preserve warmth. Blocks of buildings are much freer from smoke annoyances, because of the multiplicity of flues, which diffuse a constant warmth through the walls in which they are constructed.

There must be a sufficient supply of air flowing into the parlor to maintain vigorous combustion, else there will be defective draught. If there is a want of air, the current in the chimney will be reversed, and will flow downward instead of upward. Tightly fitting double windows, with doors listed, and weather strips at the bottom, — how can rooms thus situated receive a proper supply of air? Not only will the fire upon the hearth go out, but the unseen fires within the bosoms of the occupants of the parlor will lose their glow, and expire. One great source of smoky chimneys in city and country is the contiguity of high buildings or hills by which their tops are commanded. The smoke in such cases is beaten down by the rush of wind over them, like water over a fall. In such instances, one of two things must be done — the flue must be raised higher than the eminence, or resort must be had to somebody's patent cowl or revolving bonnet, a contrivance in such general use in cities that the lines of flues, viewed from an elevated point, look like regiments of grim warriors, with their heads dressed in ugly, fantastic gear, nodding and twirling in the wind. An immense amount of human contrivance has been expended in alterations and modifications of these appendages, as the records of our Patent Office clearly prove. And, after all, the whole matter is comprehended in the simple attachment to the flue of a rotating bonnet, so that, in whatever direction the wind blows, its mouth may be averted from it. There are chimneys which set all ingenuity at defiance, and smoke on and smoke ever, although the money expended upon them in attempts to remedy the evil may almost exceed the cost of the building of which they form an ungracious part.

Dr. Franklin, when in London, was himself thwarted in attempts to cure one of these obstinate flues. After exhausting his practised philosophy upon it, his friend, the owner of the dwelling, discovered it filled with birds' nests, upon the removal of which the evil was instantly abated.

Smoke, as we have already stated, is nothing but fuel

in a minutely subdivided state, and therefore it should be burned instead of being allowed to make its exit from the fire unconsumed. Numerous devices have been urged upon the public for the accomplishment of this object, but they are all defective in their practical workings. In large manufacturing establishments in England the burning of the smoke is common; and it would indeed be a desideratum if this result could be extended to the fires of private dwellings, as, in addition to the removal of a nuisance, there would be a considerable saving of fuel in the process. The inventive faculty can hardly be employed upon a more worthy or philanthropic object.

CHEMISTRY OF A KERNEL OF CORN.

IN considering the curious and interesting chemical nature of "corn," we shall use the term as applied to the wheat berry, as well as to the seeds of the maize plant. Among the ancients wheat was always designated as *corn;* and when we read of St. Paul's famous voyage in a "corn ship," we are to understand that the vessel was laden with Egyptian wheat. It is quite certain that neither the great apostle, nor the old Roman navigators, who held him a prisoner, ever saw a kernel of our Indian corn, — the maize plant being indigenous to the American continent.

The two grains are chemically constituted very much alike, and what may be said of one applies with almost equal correctness to the other. Both are made up of starch, dextrine, gum, sugar, gluten, albumen, phosphates of lime, magnesia, potassa, with silica and iron. Wheat contains about double the amount of lime and iron, considerable more phosphoric acid, but less magnesia and soda. Maize seeds are rich in a peculiar oil, which is nourishing, and highly conducive to the formation of adipose or fatty matter; hence the high utility of our corn in fattening animals.

What a remarkable combination of chemical substances are stored up in a kernel of corn! It may almost be said to be an apothecary shop in miniature; and the order and arrangement of the mineral elements and vegetable compounds, needed to render the comparison more apt, are not wanting. For some reason, Nature places the most valuable substances nearest the air and sunlight, while the little cells of the interior are filled full of that material used to keep erect and tidy our collars or neckbands — starch. With a moistened cloth we can rub off from the kernel about three and a half per cent. of woody or strawy material, of not much nutritive value, and then we come to a coating which holds nearly all the iron, potash, soda, lime, phosphoric acid, and the rich nitrogenous ingredients. This wrapper is the store-house, upon whose shelves are deposited the mineral and vegetable wealth of the berry. From whence come these chemical agents? By what superlative cunning were they grouped within the embrace of this covering?

They come of course from the soil, and, by the mysterious and silent power of vital force, they have been raised atom by atom from their low estate, and fitted to perform the high offices of nutrition in the animal organism. And should we not appropriate them to our use, as the most carefully adjusted of all materials designed for human aliment? Certainly we should. And do we? Unfortunately we cannot render an affirmative answer to the interrogatory. The sharp teeth of our burr mills drive ruthlessly through the rich wrapper of the kernel, and then the torn fragments pass to the bolt, and from that to the barn or stable; the animals obtain the nutritious

CHEMISTRY OF A KERNEL OF CORN.

gluten; the starch, in the form of fine flour, is set aside for household uses. But it is not designed to enlarge upon this point. Let us look at the chemical office these substances found in the kernel of corn subserve in the animal economy.

Starch is the wood or coal, which, under the influence of oxygen, is to be consumed or burned to maintain animal warmth. It passes in as pure fuel; it is oxidized, and the ashes rejected through the respiratory organs. The warmth imparted by this combustion is necessary to the proper fulfilment of the functions of the body. Of these functions, those of digestion and assimilation are the most important. The digestive apparatus receives the gluten and the starch of the grain: the latter is pushed forward to be burned; the former enters the circulation, and out of its contained iron, potash, soda, magnesia, lime, nitrogen, &c., are manufactured all the important tissues and organs of the body. All of the iron is retained in the blood, and much of the soda and phosphoric acid; the lime goes to the bones, and the magnesia abruptly leaves the body, as it seems to be very plainly told that it is not wanted. Such, in brief, are the uses which the organic and inorganic constituents of a kernel of corn subserve in the chemistry of animal life.

The changes which they are made to undergo in the laboratory are almost equally interesting and important. Fecula, or starch, is a body of great interest, and is not found alone in corn. There is scarcely a plant, or part of a plant, which does not yield more or less of this substance. What a curious vegetable is the potato! Swollen or puffed out by the enormous distention of the cellular tissue in which the starch is contained, it seems almost ugly in its deformity. It is little less than a mass of pure starch.

If we separate the starch from the gluten in corn, and boil it a few minutes with weak sulphuric acid, it undergoes a remarkable change, and becomes as fluid and limpid as water; and if we withdraw the acid, and evaporate to dryness, we have a new body, a kind of gum called "dextrine." But if we do not interrupt the boiling when it becomes thin and clear, but continue it for several hours, and then withdraw the acid by chemical means, we have remaining a sirupy liquid, very sweet to the taste, which will, if allowed to evaporate, solidify to a mass of *grape sugar*. This is the method of changing corn into sirup and sugar, about which so much has recently been said. It is a process long understood, and practically of little value. What is most extraordinary in this process is the fact that the acid *undergoes no diminution* or *change*. It is *all* withdrawn in its original amount after the experiment; nothing is absorbed from the air, and no other substance but grape sugar generated. The play of chemical affinities lies between the amidine and the elements of water, grape sugar containing more oxygen and hydrogen, compared with the quantity of carbon, than the starch.

Nothing can be more striking than these changes. From the kernel of corn we obtain starch; this we change easily into gum, and, by the aid of one of the most powerful and destructive *acids*, transform it into sirup and sugar. A pound of corn starch may thus be made over into a little more than a pound of sugar of grapes. But this result can be accomplished in another way. Let us moisten the corn, place it in a warm room, and allow it to germinate, just as do vegetables in a warm cellar. If in this condition it is dried, ground, and infused in water, a sweet liquid will be obtained, proving the presence of sugar. The change is produced, in this experiment, by the presence of *diastase*, a substance supposed to exist in malt or germinated grain, but which is imperfectly understood. The quantity of diastase necessary to effect this curious metamorphosis in corn starch is very small. We are now ready to consider another most extraordinary change which corn is capable of undergoing—that of being transformed into *whiskey* or alcohol.

If we take the sweet liquid obtained by the infusion of malted corn, and subject it to a temperature of 60° or 70° F., it soon becomes turbid and muddy, bubbles of gas are seen to rise from all parts of the liquid, the temperature rises, and there are signs of intense chemical action going on in it. After a while it slackens, and soon stops altogether. Examination shows that it has now completely lost its sweet taste, and acquired another quite distinct. An intoxicating liquid is formed, and if we place it in a still, we obtain a colorless, inflammable liquid, easily recognized as *alcohol*. By a peculiar arrangement of the condensing apparatus of the still, a portion of the grain oils and a large amount of water are allowed to go over with the alcohol; and this constitutes *whiskey*. This is an example of the change called "vinous fermentation." The influence of a ferment or decomposing azotized body upon sugar is strange, and quite incomprehensible. Through its agency, we may cause the highly organized kernel of corn to take another step downward towards a dead, inorganic condition. We can transform the alcohol over into acetic acid or vinegar, or the sugar may be formed into one of the most curious organic acids — the lactic; or, still further, it is capable of being changed into manna, a substance supposed to resemble that upon which the Israelites subsisted in the wilderness.

As in these processes we follow the kernel of corn through the various changes, first into gum, then into sugar, then alcohol, then vinegar, and ultimately into carbonic acid and water, we obtain an imperfect idea of the marvels of vital chemistry. The mysteries of these reactions have been carefully studied, and in a measure unravelled; but the necessary brief limits of this treatise will hardly allow of their consideration. The chemistry of a kernel of corn is a comprehensive topic, and to be considered even in its outlines would supply material sufficient for a volume. The aim has been to group together a few of the most interesting points, and thus awaken a desire for a more complete and satisfactory investigation.

OBSCURE SOURCES OF DISEASE.

THERE are many instances of disease brought to notice which are exceedingly perplexing in their character, and the sources of which are very imperfectly understood. They belong to a class outside of, and distinct from, the usual forms resulting from constitutional idiosyncrasies, or accidental causes, within the knowledge of the patient or medical attendant. The obscurity of their origin and persistence under medical treatment render them peculiarly trying to the patience and skill of those who have them in charge; and after the employment of the usual remedies without effect, the patients are sent into the country or to the sea-shore, as the case may be, with the expectation that a change of air or residence may prove beneficial.

We cannot, in a majority of cases, regard these affections as altogether imaginary, or as resulting from some casual derangement of the nervous system: they are instances of true disease, and should be studied with the view of bringing to light the hidden source from whence they originate. I am led to believe that a considerable number arise from some disturbance in the sanitary conditions of dwellings or their surroundings, and that, however improbable this may seem from a superficial or even careful examination of suspected premises, a still more thorough and extended search will often result in the discovery of some agent or agents capable of producing disease.

The chemical and physical condition of water used for culinary purposes has much to do with health, and is perhaps the oftenest overlooked by the physician in searching for the cause of sickness. We must not suppose that water is only hurtful when impregnated with the salts of lead or other metals; there are different sources of contamination, which produce the most serious disturbance upon the system. Some of these are very obscure and difficult of detection. The senses of taste and smell are not to be relied upon in examinations, as it often happens that water entirely unfit for use is devoid of all physical appearances calculated to awaken suspicion. It is clear, inodorous, palatable, and there is no apparent source from whence impurity may arise.

A few instances which have come under my observation may serve to illustrate the view presented, and as suggestions to those who are in doubt as regards the cause of any unusual illness.

I was recently consulted, by a gentleman residing in Roxbury, respecting the water used in his family. It was taken into the dwelling, through tin pipe, from a well in the immediate vicinity, and appeared to be perfectly pure and healthful. Analysis disclosed no salts of lead or copper, as indeed none could be expected, from the unusual precautions taken to prevent contact of the water with these metals. Abundant evidence was, however, afforded that, through some avenue, organic matters in unusual quantities were finding access to the water. Careful examination of the premises disclosed the fact that an outhouse on the grounds of a neighbor was so situated as to act as a receptacle for house drainings, and from thence, by subterranean passages, the liquids flowed into the well. Some cases of illness, of long standing in the family, disappeared upon abandoning the use of the water.

A specimen of water was brought to me, for chemical examination, by a gentleman of Charlestown, who stated that his wife was afflicted with protracted illness of a somewhat unusual character. It was found to be largely impregnated with potash and the salts resulting from the decomposition of animal and vegetable *debris*, and the opinion expressed that some connection existed between the well and the waste fluids of the dwelling. This seemed improbable, as all these were securely carried away in a brick cemented drain, and in a direction opposite the water supply. The use of the spade, however, revealed a break in the drain at a point favorable for an inflowing into the well, and hence the source of the contamination. Rapid convalescence followed on the part of the sick wife upon obtaining water from another source.

Analysis was recently made of water from a well in Middlesex County, which disclosed conditions quite similar to these. The owner was certain that no impurity could arise from sources suggested; but rigid and persistent investigation disclosed the fact, that the servant girl had long been in the habit of emptying the "slops" into a cavity by the kitchen door (formed by the displacement of several bricks in the pavement), where they were readily absorbed. Although the well was quite remote, the intervening space was filled with coarse sand and rubble stones, and hence the unclean liquids found an easy passage to the water. This proved to be the cause of illness in the family.

It is unnecessary to present other instances of a similar character on record. These serve to bring to view some of the sources of impurities in water used for household purposes, and the obscure cause of serious diseases. The location of wells connected with dwellings is a matter which should receive careful attention.

It is well known that in the gradual decomposition of animal and vegetable substances, at or near the surface of the earth, under certain conditions, nitrogenous compounds are developed. The nitre earths found beneath old buildings result from these changes, although it is quite difficult to understand the precise nature of the chemical transformations which produce them. In the waters of a large number of wells in towns and cities, and also in the country, the nitrates are found at some seasons in considerable quantities. The salts form at the surface in warm weather, and, being quite soluble, are carried with the percolating rain water into the well. In cities and large towns, where excrementitious matters accumulate rapidly around dwellings compacted together, it is difficult to locate wells remote from danger; and hence it might seem that suspicion should be confined to these localities. This, however, is not a safe conclusion. How often do we see, upon isolated farms in the country, the well located within, or upon, the margin of the barnyard, near huge manure heaps, reeking with ammoniacal and other gases, the prolific sources of soluble salts, which find access to the water, and render it unfit as a beverage for man or beast. It may, no doubt, be a convenience to the farmer to have his water supply so situated as to meet the wants of the occupants of his barn and his dwelling; but it is full of danger.

Whilst admitting that such may be the condition of the water of many wells, doubts may arise, with some, whether substances not decidedly poisonous, and received in such quantities, can, after all, be productive of much harm, or the real source of illness. To a large number of people they are certainly harmless; but it must be admitted that there is a class—and one or more are found in almost every family—whose peculiarly sensitive organization does not admit of the presence of any extraneous agent in food or drink, or in what they inhale. The functions of life and health are disturbed by the slightest deviation from the usual or normal condition of things around them. It is manifestly of importance that we should recognize these peculiarities in individuals. It is unsafe, in making a diagnosis of disease, or seeking for causes, to overlook or forget them.

We are, indeed, incapable of understanding how this can be. It seems incredible that the thousandth part of a grain of one of the salts of lead, dissolved in water and taken daily, will disturb the system of any one; and yet such is the case. We can see no reason why a very little nitrate of potassa, or soda, or lime, taken in the same way, should produce any ill effects; still stranger is it that the infinitesimal amount of dust dislodged from painted wall-papers, received into the lungs, should make inroads upon health.

Several instances of this latter result have recently come to my knowledge. In two families of the highest respectability in this city, illness of an unusual and protracted character existed, and at the suggestion of the physician, portions of the green wall-paper of the dwelling were submitted to me for analysis. The pigments were found to consist mainly of arseniate of copper, and upon the removal of the papers the illness disappeared. In experimenting with apparently the most suitable apparatus, and employing delicate chemical tests, in rooms the walls of which were covered with these arsenical papers, no evidence of the presence of the poison in the atmosphere has been afforded; and this corresponds with the results of all similar experiments made in this country and in Europe, so far as my knowledge extends. We must conclude that agents not recognizable by chemical tests are capable of disturbing vital processes. The evidence is very clear that in instances of illness confined to one or two members of a household, the cause may be due to some accidental disturbance with which all are equally brought in contact, but which has the power of injuriously influencing but a part. It is also clear that these sources of disease are of such a character as easily to escape detection, and therefore any facts or experience which may serve as guides to their discovery, are worthy of consideration.

LOCAL DECOMPOSITION IN LEAD AQUEDUCT PIPES.

IN most cities and towns supplied with aqueduct water, through leaden pipes, a confident feeling prevails that the general influence of the water is harmless, and no one suspecting any decompositions to result from *local* causes, the idea of lead-poisoning does not enter the mind of any consumer.

Competent chemists are commissioned to make careful and extended experiments to ascertain the effect of the water upon lead, and their reports generally assert the the non-liability of contamination. It is just that confidence should be reposed in the results of their investigations. The experiments and conclusions, respecting the *general* influence of the waters upon lead, are usually accurate and reliable.

There can be no doubt that the waters of Cochituate Lake, like those of most New England ponds, in their freedom from chlorides and nitrates, and generally holding in solution sufficient carbonic acid to change soluble oxides into insoluble carbonates, are safe to use after passing through lead pipe, under ordinary circumstances

But to form an opinion of their entire safety at all points of delivery, we must inquire if the relationship of chemical forces may not be so affected or changed in one locality, as to change the character of the water flowing in that direction. We certainly ought to infer that such is the fact, when the presence of lead is detected in the water, and cases of lead disease are found following its use.

Several years ago, the writer called attention to the instances and causes of local decomposition in lead pipes, through a public journal, and since that time the additional instances that have come to his knowledge have convinced him of the importance of the subject.

The late Dr. Treadwell, of Salem, several years since, suspected that he was suffering from lead disease, and sent to me, for analysis, samples of water supplied to his dwelling. The amount of the metal present was found to be large; so large, that, for the purpose of obtaining a comparison of results, a portion was sent to a distinguished chemical friend for examination. The results in no respects differed. The violence of the symptoms in Dr. Treadwell's case rapidly abated upon abstaining from the use of the water. A specimen of this aqueduct water, taken from another locality, afforded a trace of lead, while that from other pipes gave no lead reaction with the most delicate tests.

Instances of the kind, that have come under my observation, and those on record, are numerous. It is safe to say that there is no time when there are not individuals in all cities and towns suffering from lead disease. It is marvellous how susceptible some individuals are to the influence of this metal in the system. I have been made acquainted with a case where two members of a family of seven were made seriously ill from the use of water containing only, at times, a mere trace of lead — a quantity so infinitesimally small as not to have the least effect upon the health of the others.

In view of the facts, it seems necessary to inquire, What produces this lead impregnation in certain houses or districts, while the general waters of a supply remain unaffected?

In the course of investigations several interesting facts have been developed, tending to throw light upon this subject. I have noticed in the leaden pipes removed from cess-pools, sinks, and wells, that the intensity of corrosive action had been in a great measure confined to the sharpest bends and depressions in the pipe, and in some instances other portions remained intact.

I have in my possession a section of supply-pipe, removed from the aqueduct of a neighboring city, in a portion of which corrosive action had proceeded so far as to cause leakage. The part thus acted upon was confined to an acute angle, and there is evidence to show that the plumber, in placing it in position, bent it in the wrong direction, thus creating the necessity for another turn in the opposite. This pipe had doubtless been subjected to two violent turns, which seriously impaired the homogeneity of the metal. An examination of lead pipe removed from buildings will certainly show that where there has been any perceptible amount of decomposition, it has been confined to the angles and depressions in its course.

There are three causes or agencies which may, perhaps, be sufficient to produce these results: —

1. The disturbance in the crystalline structure of the metal by bending, whereby its electrical condition is changed, and voltaic action promoted, giving rise to chemical decomposition.

2. The presence of organic matter, such as fragments of leaves, and impurities pervading all pond waters, and which may be detained in angles and depressions of the pipes. Their presence, undoubtedly, promotes oxidation, and the protoxide of lead will remain in solution, unless sufficient carbonic acid is furnished to change it. It is easy to conceive of conditions where this could not be the case.

3. Corrosions may be produced in lead pipes by the accidental presence of pieces of mortar. Where mortar is present, the lime would assist in oxidizing the metal, and also aid in the solution of the oxide. Considerable portions of fresh mortar are frequently deposited in lead pipes during the erection of buildings. When the family commence the use of the water, it holds the salts of lead in solution, and its presence may be detected for months. The process of oxidation, which is retarded or prevented altogether by the presence of neutral salts in water, could not be materially interfered with under the conditions considered.

It is obvious, if these observations and conclusions are correct, that much care should be exercised in placing pipes in position in buildings. In those leading to the culinary department, angles and depressions should be avoided. Violent twists and turns should not be permitted; and during the erection of houses, the open ends of protruding pipes should be carefully closed.

Assuming the general fact that lead pipes, conveying the waters of our New England ponds, become coated and protected by an insoluble lead salt, the question arises, How long before this protection is secured? or, How soon may a family commence the use of water passing through new pipes, with safety? In view of the manifest danger from local disturbances, the most sensible reply would be, *Never.* A section of new lead pipe, immersed in Cochituate water one hour, at a temperature of 65° Fahr., gave a decided lead reaction with sulphydric acid. Removed, and placed in six fresh portions of water, one hour in each, the waters, when tested, gave similar results. The experiment continued during two weeks. Varying the time of immersion in fresh portions of water from one hour to ten, the lead indications continued, although at last feeble. These results are sufficient to show that individuals or families should not commence the use of water flowing through new pipes until a considerable time has elapsed, and much water contact secured.

It is important that medical gentlemen should be made fully aware of every source from whence disease may arise; and if there are symptoms in patients indicating lead affections, it would seem desirable that investigations should be instituted to ascertain the facts, although there may be no apparent source through which the salts of lead could be introduced into the system.

BREAD AND BREAD-MAKING.

FROM the character of bread offered for premiums at the exhibition of agricultural societies the conclusion is reached, that very many families have hardly yet learned what good bread is, and that there is a wide margin for improvements in the methods of bread-making. No subject is certainly more important, as it has a direct bearing upon the health and consequent happiness of households, and it should receive the attention which it deserves.

Besides the manipulating processes, the manufacture of good bread involves some other considerations of no secondary importance. It is useless to attempt its production with imperfect or bad materials. The flour or meal must be sweet, and from fully matured grain. During every year the market is crowded with flour of a damaged character. Severe rains and long-continued moist weather, which prevail at the South and West, are unfavorable for securing the grain crops, and much of it germinates in the fields and barns, and is thereby rendered unfit for bread-making. In the germinating process, disastase is formed; this, reacting upon the starch of the flour in the baking, transforms it into dextrine and sugar, and prevents the formation of light, spongy bread. Flour from such grain will afford only sticky, glutinous, heavy bread, no matter how much care and skill is bestowed in the making. Fungous growths also appear in wheat injured by moisture, and the flour becomes "musty." In bread from such materials, beside its repulsive physical appearance and unpleasant taste, a chemical change has occurred which renders it positively injurious as an article of diet. The nutritive properties, the gluten, especially, has undergone decomposition, and new bodies have been formed, which are not of an alimentary nature. Impaired digestion, derangements of the bowels, follow the use of bread from such flour. The poor, who are unable to pay large prices for choice, selected brands, suffer greatly from this source, and much of the bread they are compelled to eat is well calculated to weaken rather than sustain the vital functions.

During the most favorable seasons thousands of bushels of wheat are made into flour, which, owing to local causes, delay in harvesting, or storage in large bodies, is rendered entirely unfit to be used as food. A portion of this is employed in the arts; but the great bulk goes into families, and feeble children, as well as adults, are forced to consume it, much to their injury. It is doubtful if anything can be done to abate this evil; the cupidity of men is but little affected by considerations of right, and the thirst for gain is potent and irresistible.

There are several methods of testing wheat flour, which are available to purchasers, although none of them afford positive indications. Good flour is not sensibly *sweet* to the taste, but bad flour often is. This is owing to the presence of glucose, resulting from chemical changes in the grain, by partial malting. Extreme whiteness is a good indication, as changed grain is discolored in the process of change. Good flour is tenacious and unctuous to the touch; when thrown against a wall, it should adhere, and not fall readily. It does not feel *crispy*, and when formed into a ball in the hand, adheres together like a ball of snow. To the sense of smell it is sweet and pleasant, and when taken into the mouth, forms a glutinous mass, free from all disagreeable taste.

The nutritive quality of flour depends upon the proportion of gluten which it contains. In the best specimens ten or twelve per cent. is found. A barrel of flour contains about twenty pounds of gluten, and one hundred and fifty of common starch. The starch can easily be washed out of a small quantity of flour by placing it in a bag of cotton cloth and kneading it under a stream of water. The gluten remains upon the cloth, and is a gray, viscid, tenacious mass, insoluble in water. It is the strength-giving principle of the flour, and in a three-pound loaf of bread there should be at least three ounces of this substance.

Bad bread is by no means always chargeable to imperfect materials. Hundreds of families, who procure and use the most perfect flour, subsist upon bread of a very inferior quality. Some housekeepers assert that they can have no "luck" in bread-making; their loaves are always heavy, or sour, or doughy, or burnt, and they give up experimenting and become discouraged. As with good materials every one can prepare good bread, there should be no want of success.

Success depends in a great measure upon good judgment, faithfulness and patience in working, and in using the right materials. It is quite preposterous to present a fixed recipe and set it up as an infallible guide in this department of household labor. The method adopted in my family, which affords perfect white bread, is as follows: —

Sift five pounds of good flour and put it in an earthen pan suitable for mixing and kneading. Have ready a ferment, or yeast, prepared as follows: —

Take two potatoes the size of the fist, boil them, mash and mix with half a pint of boiling water. A fresh yeast cake, of the size common in the market, is dissolved in water, and the two solutions mixed together and put in a warm place to ferment. As soon as it commences to *rise*, or ferment, which requires a longer or shorter time, as the weather is warm or cold, pour it into the flour, and with the addition of a pint each of milk and water, form a dough, and knead for a full half hour. Form the dough at night, and allow it to stand until morning in a moderately warm place; then mould and put in pans, and let it remain until it has become well raised; then place in a hot oven and bake.

The points needing attention in this process are several. *First*, the flour must be of the best quality; *second*, the potatoes should be sound and mealy; *third*, the yeast cake is to be freshly prepared; *fourth*, the ferment must be in just the right condition; *fifth*, the kneading should be thorough and effective; *sixth*, the ra sing of the dough must be watched, that it does not proceed, too far and set up the acetic fermentation, and cause the bread to sour; *seventh*, after the dough is placed in pans, it should be allowed to rise, or puff up, before placing in the oven; *eighth*, the temperature of the oven, and the time consumed in baking, have much to do with the perfection of the process.

If this method is followed with the exercise of good judgment and ordinary skill, white bread of the highest perfection will be uniformly produced.

Unfermented, or "cream of tartar" bread, is never placed upon the table in my family. There are special dietary or sanitary reasons for its exclusion. All "quick-made" bread is usually prepared in haste, and the adjustment of acid and alkali is apt to be imperfect. Not one pound in a hundred of cream of tartar sold in the market is free from adulteration. In ten specimens procured from as many different dealers, in a town of ten thousand inhabitants, I ascertained by analysis that the *least* percentage of adulterating material was twenty-two per cent., and several were over seventy per cent. The "yeast powders" so common in the market are composed of acids in association with alkaline carbonates, usually bi-carbonate of soda. If tartaric acid, or cream of tartar, is used with the soda, there remains in the bread after baking a neutral salt, the tartrate of soda, which is diffused through the loaf and is consumed with it. This salt has aperient properties — in fact, is a medicine; and thus, at the daily meal, those who use bread made with "powders," or with cream of tartar, are taking food and medicine together.

Some years ago, Professor Horsford, of Cambridge, proposed substituting phosphoric acid for the tartaric; and this excellent idea has been put into practical effect in the production of yeast powders. In the use of this acid, *phosphate* of soda would remain in the loaf; and as this is made up of the element which we lose in sifting out the bran from the flour, it must prove healthful, or at least unobjectionable. But bread prepared by *effervescing* powders is, at best, a poor substitute for that which results when the dough is raised through the agency of vinous fermentation — regular yeast, in some of its forms, being employed. Effervescents may be used in exigencies, which occasionally occur; but it is hoped that the good housewives in our country do not, in their bread-making, habitually depart from the good old way of raising the loaf by panary fermentation.

It was a noticeable fact that seldom specimens of whole meal, wheaten, or corn bread are offered for exhibition. It is presumed that the premiums of agricultural societies are intended to include these forms of the "staff of life," and it is a matter of regret that none are presented. There is manifestly a perversion of sentiment, or fashion, as regards bread made from the unbolted meal of wheat, which ought to be corrected. Why, upon the tables of farmers, the white flour loaf should usurp the place of the darker, but sweeter and more healthful one from the whole meal, is a question of no little interest and importance. In the Eastern States, but few soil cultivators raise this noble grain in quantities large enough to meet family wants; and it is probable, if the reverse of this were true, the grist would be carried long distances to a mill with a bolt, to separate the fine flour.

If there is any one form of bread more delicious than another, or more conducive to the sustentation of the physical and intellectual powers, it is that from unsifted wheat meal; and every owner of land should include this grain among his crops, that he may have the bread fresh and in its highest perfection. A generous dressing of finely ground bone will put almost any field in condition to grow a profitable crop; and in these days, when flour of the better sorts commands such enormous prices, there seems to be no good reason why farmers should not resume the cultivation of wheat in all wheat-growing states.

Corn bread is also excellent and most nutritious. It contains a large amount of oil not found in other grains, which adds greatly to its value. There is far too little of this used in our families. The old-fashioned dish of corn "pudding and milk" is now nearly as obsolete as that of "bean porridge;" and may we not, with much reason, attribute the physical degeneracy of the present race to the radical changes in the forms of food? Regarding the matter from a chemical and medical point of view, it certainly would be difficult to select better or more healthful forms of human nutriment — forms so well calculated to build up and sustain a "sound mind in a sound body," as the two named above, once so popular, but now banished from our tables. They were easy of digestion and assimilation, and contained all the chemical substances, or organic and inorganic constituents needed to nourish the body and mind. Certainly, white flour bread, cake, and condiments are poor substitutes for the sensible but plain dishes of our fathers and mothers a half century ago.

CHEMISTRY OF THE SUN.

IT will be quite natural for the reader, not acquainted with the triumphs of all-conquering Science, to exclaim, as he casts his eye upon the title of this essay, "Chemistry of the Sun! How can we know anything of the chemistry of a body so remote? With the enormous distance of ninety-five millions of miles intervening between our mundane chemists and the fiery orb, how can they obtain knowledge regarding its structure, or the materials of which it is composed?" Incredible as may seem the fact, they have acquired knowledge upon these points.

Chemists are certain that several of the metals common to our earth exist upon the sun, or form a part of that incandescent covering from which proceed light and heat. This knowledge rests upon experimental demonstration. However insignificant is man, and confined to a narrow belt upon this little ball of earth, he is proved capable of stretching his powers through unlimited space, and estimating the chemical composition not only of the sun, but of the fixed stars, and, by these sublime researches, affording evidence that there is entire unity of matter throughout the vast physical universe.

Let us proceed to the examination and explanation of the wonderful nature of that analytical method by which these results are reached. Of course it must be conceded that the only channel through which we can gain any knowledge of the distant heavenly bodies, is through the vivifying radiance which they so constantly pour forth into surrounding space, and which impinges upon our planet after considerable time is consumed in its rapid flight. It is through the agency of imponderable *light* that all our chemical knowledge of the sun is derived. The scientific world is indebted to Kirchhoff, Professor of Physics in the University of Heidelberg, for the fullest and most accurate explorations in the field of *spectrum analysis*—a method of demonstration by which light is proved to be a trusty messenger for bringing us tidings regarding the chemical nature of distant worlds. In connection with Professor Bunsen, he made a series of experiments, which resulted in the establishment of a new method of chemical analysis of marvellous delicacy and accuracy, based upon the peculiar effects of light given off by terrestrial matter, when through heat it becomes luminous. This method is proved to be so delicate as to enable the analyst to detect with ease and certainty so minute a quantity as the $\frac{1}{180,000,000}$ part of a grain of any substance.

It is somewhat difficult to convey a very clear idea of this method to those not acquainted with optics or chemical science; and in this discussion we shall enter only so far into the matter, as to afford a general view of the philosophy of the process, and the nature of the apparatus employed.

If we allow a beam of sunlight to pass through a round hole in the shutter of a darkened room, and interpose a triangular piece of glass, called a prism, so as to cause the ray to traverse it at a peculiar angle, we find, instead of a white spot of light upon the wall, a bright band of various colored lights, showing all the tints of the rainbow. The white light is split up into various constituent parts, thus proving that what we call *whiteness* is composed of an infinite number of differently colored rays. This experiment by no means reveals all the beauties of solar light, as the colored lights overlap and interfere with each other, and the colors are confused. If, for the round hole, we substitute a narrow slit in the shutter, and allow the ray to pass through the prism, we shall notice that a number of *dark lines* appear, cutting up the colored portion of the spectrum, and essentially modifying its appearance. What is the meaning of these dark lines? What do they indicate? In Newton's time they were thought to have no meaning. Dr. Wollaston, in 1802, counted them, and could distinguish only *seven;* and to him they were without meaning. To modern chemists these dark lines are full of the deepest interest, and upon them an immense amount of study has been bestowed.

Fraunhofer, in 1815, by improved optical arrangements, had increased Wollaston's *seven* dark lines to five hundred and ninety; and a few years later, Sir David Brewster mapped with great care nearly two thousand in the portion of the spectrum from red to violet. Thus rapidly does persistent scientific investigation unfold the mysteries of nature, and extort secrets which for years seemed hidden in impenetrable darkness.

Professor Kirchhoff, in his experiments, employed four prisms, and the workmanship was of the most perfect kind. The dark lines upon the spectrum were observed through a telescope having a magnifying power of forty; and thus he was enabled to attain a minute distinctness of observation which far excelled that of all previous observers. He saw vast numbers of dark lines, and with marvellous skill he measured their relative distances, and mapped them, as the geographer maps rivers and islands, or as the surveyor triangulates and marks out the main features of a country. Splendid indeed are the results of the researches of the German philosopher, and his name is imperishable.

But it is time to inquire how light reveals the constitution of earthy materials, or how it can afford us information regarding the chemical nature of the solar atmosphere. Every one who has witnessed a pyrotechnical display, remembers the brilliant red and green fires which so dazzled the sight, and excited the admiration. The beautiful lights were produced by burning the salts of certain metals: the crimson came from the salts of strontia; baryta and copper gave *green;* soda, *yellow;* potassa, *white,* &c. It has long been known that these metals, or their salts, afforded certain peculiar colors when burned at a high heat; but it was not known until recently, that all metals and all elementary substances emit, when heated, a characteristic kind of light — a light peculiar to each element. By this discovery, the great principle of spectrum analysis was revealed; and to it we are indebted for all the sublime revelations which have flowed from this new method of research.

Every element, whether it be a gas, a solid, or a liquid, when heated, emits a light, so distinct, so peculiar, that we have only to see and examine the light, to know what particular element is being heated. It follows, if we have sixty or more elementary bodies, we must have that number of different impressions produced upon the visual organs when they are ignited or heated. If we are possessed of a perfect knowledge of these peculiarities, and understand how to arrange substances for observation, it is clear we must have a means of analysis, or of learning of what bodies are composed, entirely independent of the use of test tubes, blow-pipes, or reagents of any description. It is only in the gaseous state that each kind of matter emits the light peculiar to itself. How can solids be made gaseous? How can gases be heated so as to become luminous? Every substance can be volatilized and made gaseous, no matter how refractory it may be; and all flame is nothing else than heated gas. By the lighting of a candle we establish miniature gas works. The solid wax or tallow is changed into gas, and the gas is heated until it becomes luminous, and affords light. By other methods, gold, silver, copper, iron, &c., can be changed to the aeriform state, and burned.

If we wish to examine the flame of the metals of the alkalies, as sodium or potassium, we have only to take one of their salts (the chloride), and bring a small quantity into the flame of a spirit lamp. It immediately volatilizes, and is heated so as to tinge the flame with its peculiar color. In order to become acquainted with the exact nature of the light which bodies in the condition of luminous gases emit, their light must be examined otherwise than by the naked eye. We must examine the *spectra* which the light affords; and to this end, an instrument corresponding with that used for examining the solar ray is employed. The light from the heated substance is passed through a slit in a tube, and is refracted by prisms, and the spectra viewed by the aid of a high magnifying power. The instrument is called a *spectroscope*.

We remember the surpassing beauty of the spectrum of a ray of sunlight: let us notice the difference afforded by a ray proceeding from heated sodium. To obtain this light, we throw into the flame of an alcohol lamp a grain or two of common salt (chloride of sodium), and place it before the slit in the tube. Now, by looking through the telescope we observe the spectrum of sodium; and what a contrast it presents to that of sunlight! It consists of two very fine bright yellow lines placed close together: *all the rest of the field is perfectly dark.* No other substance in nature affords a spectrum like this: when it is seen, sodium in some form is the producer. If we had employed the chloride of potassium instead of the soda, the spectrum would have afforded a portion of continuous light in the centre, bounded by a bright red and a bright violet line at either end. Other metals would in like manner have afforded their peculiar and distinct spectra. None of the bright lines overlap, or interfere with each other; and if we should put into the flame chlorides of all the metals, if it were possible to obtain them, all the different characteristic lines would come out perfectly clear and distinct.

From these statements, it follows that if the light of the sun was produced entirely by heated sodium, the spectra would afford only the two yellow lines, and dark lines would cover all the rest. The dark lines indicate the *absence* of other kinds of rays. They are shadows, as it were, in the background. If two, or three, or a dozen elements, similar to those which enter into the construction of our planet, were being volatilized and heated at the great central source of solar light and heat, characteristic spectra would be afforded, indicating their presence. Bright lines would indicate the presence, the dark lines the absence, of certain substances in direct sunlight.

We are now prepared to understand in a general way the nature of the new method of analysis by which sunlight, and light from every source, is made to reveal the character or chemical composition of the body from which it emanates.

Before proceeding to remark further upon this department of the subject, let us consider briefly the physical character of the sun. The science of solar physics is not far enough advanced to give us much positive knowledge regarding its physical structure. Research has, however, rendered very nearly certain one point of much interest. This relates to the *condition* of the matter from whence proceed the solar light and heat. There are two strong experimental proofs that actual combustion is going on at the sun, or some process analogous to it, and that matter exists there in the condition of flame. The first of these proofs is afforded by the nature of the light emitted. All bodies rendered incandescent by heat, which are in the liquid or solid state, emit invariably *polarized* light; while bodies which are gaseous, when rendered incandescent, emit light in its natural state, or unpolarized. Thus the physical condition of a body may be distinguished, when incandescent, by examining the light it affords. On applying these tests to the direct light of the sun, it is found to be in the unpolarized or natural condition; hence the matter from which the light proceeds must be in the gaseous state. The character of the solar spectrum affords another or the second form of proof that flame exists at the sun. If we heat a metallic body to a white heat, and bring it in connection with a spectroscope, and examine its light, we shall find that the spectrum is crossed by *no dark lines* — they are all absent. If we examine any *solid* or *liquid* incandescent body, the dark lines are wanting in the spectrum. Not so with the light from incandescent gaseous bodies, or from actual flame; this invariably furnishes the dark lines in greater or less abundance. The solar spectrum, as we have seen, is crossed by thousands of dark lines; hence the light must proceed from volatilized or gaseous matter intensely heated.

It is singular that Kirchhoff adopts the same theory regarding the physical constitution of the sun as that of Galileo, who wrote so many ages since. It should be said, however, that this theory has received but few essential modifications from modern science. To explain it, it is necessary to allude to the dark spots which are so often seen upon the sun's disk, and which have excited most intense interest among astronomers for centuries. These spots, viewed through a telescope of high power, appear to be an intensely black, irregularly-shaped patch, edged with a penumbral fringe, the brightness of the general surface of the sun gradually fading away into the blackness of the spot. Dr. Peters, while connected with the Observatory at Naples, computed eight hundred and thirteen heliographical places of two hundred and eighty-six spots. His paper, read at the Providence meeting of the American Association for the Advancement of Science, is an exceedingly interesting one. The spots upon the sun have a motion of their own, and they appear and disappear in a most singular manner. New spots generally break out to the east of old ones, and have a motion towards the west, and the motions in longitude are far more considerable than in latitude. These movements are, in some instances, at the rate of four hundred miles an hour. Two zones upon the sun's surface are particularly fruitful in spots, the maximums occurring at the parallels of 21° of north latitude and 17° of south. When a spot is about to appear, a sort of bubbling agitation in the luminous mass is seen; a small, dark point is rapidly developed, and in the course of a day the full area of the spot is attained. They continue of full size sometimes thirty or even fifty days, and then notches begin to form in their margins; these grow larger and deeper, until, by a sort of electric flash, the realms of light close up, and the spot is gone. The magnitude of these spots is in proportion to the magnitude of the sun itself. An object ninety-five millions of miles distant, to be barely visible, must have a diameter of four hundred and sixty miles, and an area of one hundred and sixty-six thousand square miles. Therefore the smallest spot obse ved must be of this immense size.

Among the many spots recorded, one by Mayer had an area of *fifteen hundred millions of square miles*, or about thirty times the surface of the earth.

What are the spots? Galileo described them as being clouds floating in the gaseous atmosphere of the sun, appearing to us as dark spots on the bright body of the luminary. Sir William Herschel supposes that the centre of the spot reveals a portion of the dark surface of the sun, seen through two overlying openings, one formed in a photosphere, or luminous atmosphere, surrounding the dark, solid nucleus, and the other in a lower opaque or reflecting atmosphere.

Kirchhoff says that the most probable supposition which can be made respecting the sun's constitution is, that it consists of a solid or liquid nucleus, heated to a temperature of the brightest whiteness, surrounded by an atmosphere of somewhat lower temperature.

What causes the spots? Dr. Peters regards them as caused by volcanoes upon the solid mass of the sun. These volcanoes, in a state of high activity, would send up vast volumes of gaseous matter, which, by its disruptive force, would part the luminous covering, and expose to view the opaque body within.

M. De la Rue, in connection with Balfour Stewart, Esq., and Benjamin Loewy, Esq., of the Kew Observatory, England, has very recently published a memoir for private circulation, in which the ground is taken that they are caused, or greatly influenced, by external disturbing forces, and that the planets Venus and Jupiter have much to do with the behavior of the spots. Venus, being very near the sun, influences it more than any other body, and causes vast mechanical changes. The molecular state of the sun is exceedingly sensitive, the least withdrawal of heat causing condensation or cooling, and this altered state of the photosphere appearing like dark patches upon the sun's disk. The reasoning, supported as it is by a series of patient observations, calculations, and experiments, is certainly worthy of high regard and confidence.

If this view is correct, why may not our seasons be greatly modified by the withdrawal of solar heat, by planetary influences? And again, why may not the moon, being a much cooler body than the earth, influence the seasons and the weather upon our planet? We evidently have yet much to learn regarding these questions in physical science.

It has been conceived possible, by some philosophers, that the sun may be peopled by beings essentially like ourselves, the opaque mass upon which they are placed being protected from the intense light and heat of the luminous covering by some non-conducting media, adapted to the purpose. According to Kirchhoff, this notion is wholly untenable, as he undertakes to show that the idea of an intensely ignited photosphere, surrounding a cold nucleus, is a physical absurdity. As interesting speculations, the different views of different men are worthy of notice.

The opinion of all observers seems to coincide in

regard to the nature of the outer coverings of the sun, that it must be matter in a highly heated condition; the state of the central mass is a question about which there is a conflict of views.

From a careful consideration of all experimental observations recorded, it seems highly probable that the sun is an incandescent mass; that it is a vast globe of matter in a state of ignition, fusion, or volatilization. The idea that any process, analogous to that of common combustion, is going on there, seems to be beset with formidable difficulties. The researches, however, of Mayer, Grove, Helmholtz, and Tyndall tend to remove one of the most serious obstacles in the way of entertaining this belief; and that relates to a supply of material to support so enormous and interminable combustion. The amount of heat emitted by the sun every minute, is competent to boil twelve thousand millions of cubic miles of ice-cold water. No combustion, no chemical affinity with which we are acquainted, would be competent to produce such a temperature. The heat is radiated into space, and out of 2.300 millions of parts sent forth, the earth receives *only one part*. It is possible that asteroids, or meteoric masses, falling into the sun, may furnish a supply of material for making good the immense loss of heat. There is every reason to believe that space is stocked with these bodies. That peculiar haze, or envelope, which surrounds the sun, called the zodiacal light, is probably a crowd of meteors; and moving as they do in a resisting medium, they must approach the sun and rain into it with irresistible force. This would constitute a source from which the annual loss of heat might be made good.

The instant these bodies touch the focus of heat, they must be volatilized, and themselves afford light and heat. Whatever might be their composition, the atoms would exist in a form similar to the sodium in the flame of the alcohol lamp, spoken of in the first part of this essay. They would be in condition suitable for analysis by means of the spectroscope. Although the incandescence of these bodies takes place at such an immense distance from our field of observation, yet the light they emit reaches us, and by means of the spectra afforded, we may learn something regarding their chemical composition. If the rays of light which form the solar spectrum have been filtered through the vapors of iron, or nickel, or copper, existing at the sun, these metals will manifest themselves there. And do they? They certainly do. The spectrum of sunlight shows the presence of iron, chromium, nickel, cobalt, barium, copper, zinc, &c. — the same metals which are common to our earth. This result is no myth, no vague supposition, no possible contingency. We *do know* something of the chemical nature of the sun. A deeper penetration is made into the mysteries of the far distant spheres; we have moved one step farther in that onward march of science which seems destined to overcome all obstacles, and bring us into intimate communion with the vast universe of God.